Machine Interpretation of Line Drawing Images

Springer

London
Berlin
Heidelberg
New York
Barcelona
Hong Kong
Milan
Paris
Singapore
Tokyo

Sergey Ablameyko and Tony Pridmore

Machine Interpretation of Line Drawing Images

Technical Drawings, Maps and Diagrams

Sergey Ablameyko, PhD, DSc, Prof, FIEE, FIAPR, SMIEEE
Institute of Engineering Cybernetics, Belarusian Academy of Sciences, Minsk, Belarus

Tony Pridmore, BSc, PhD
School of Computer Science and Information Technology, University of Nottingham, Nottingham, UK

ISBN 3-540-76207-8 Springer-Verlag London Berlin Heidelberg

British Library Cataloguing in Publication Data
Ablameyko, Sergey
  Machine interpretation of line drawing images : technical
  drawing, maps and diagrams
  1.Optical pattern recognition 2.Image processing - Digital
  techniques
  I.Title II.Pridmore, Tony
  621.3'67
  ISBN 3540762078

Library of Congress Cataloging-in-Publication Data
Ablameyko, Sergey, 1956-
    Machine interpretation of line drawing images : technical
  Drawings, maps and diagrams / Sergey Ablameyko and Tony Pridmore.
        p.   cm.
    Includes bibliographical references.
    ISBN 3-540-76207-8 (alk. paper)
    1. Image processing--Industrial applications. 2. Document imaging
  systems. 3. Optical scanners. 4. Optical pattern recognition.
  I. Pridmore, Tony, 1961- . II. Title.
  TA1637.A26 1999
  621.36'7--dc21   99-17199

Apart from any fair dealing for the purposes of research or private study, or criticism or review, as permitted under the Copyright, Designs and Patents Act 1988, this publication may only be reproduced, stored or transmitted, in any form or by any means, with the prior permission in writing of the publishers, or in the case of reprographic reproduction in accordance with the terms of licences issued by the Copyright Licensing Agency. Enquiries concerning reproduction outside those terms should be sent to the publishers.

© Springer-Verlag London Limited 2000
Printed in Great Britain

The use of registered names, trademarks etc. in this publication does not imply, even in the absence of a specific statement, that such names are exempt from the relevant laws and regulations and therefore free for general use.

The publisher makes no representation, express or implied, with regard to the accuracy of the information contained in this book and cannot accept any legal responsibility or liability for any errors or omissions that may be made.

Typesetting: Camera ready by authors
Printed and bound at the Athenæum Press Ltd., Gateshead, Tyne and Wear
34/3830-543210 Printed on acid-free paper SPIN 10644759

Preface

This book is concerned with the theory and practice of computer interpretation of images of line drawings. At present, many companies, agencies and individuals throughout the world regularly invest significant financial and human resource in the input of information into computers. Much of the information involved is presented in the form of line drawings. Manual input of those drawings to, for example, a Computer-aided Design (CAD) or Geographical Information System (GIS) is a laborious and time-consuming task. Line drawings are often large, containing many narrow, closely spaced line segments, each of which must be separately identified in the computer-based representation. Manual input generally involves tracing over the drawing with an appropriate pointing device, an operation that produces at best variable and often erroneous results. The availability of automatic and/or easily usable, interactive line drawing input systems would make this important process both easier and cheaper.

The technological infrastructure needed to support the development of such systems is already in place. Equipment capable of efficiently scanning a document to produce a digital image is now widely available, as is software capable of basic image manipulation, storage and display. The extraction from the resulting image of the representations required by target applications such as CAD and GISs is, however, a complex problem that remains an active area of research and development.

Line drawings arise in a wide variety of disciplines and situations. To provide a focus for discussion, we will deal primarily with the transformation of images of engineering drawings and maps into formats required by mechanical/manufacturing engineers and cartographers respectively. Many of the issues discussed and techniques described are, however, relevant to the interpretation of other types of line drawing. One might wish to consider, for example, images of floor plans, electrical schematics or data flow diagrams. As in these other domains, machine interpretation of maps and engineering drawings involves recognition of quite abstract graphical objects. Engineering drawings might be described in terms of, e.g. the outlines of physical objects, cross-hatched areas, centre lines and dimension sets. Entities of similar complexity (contour lines, roads, pipelines and cables, etc.) must be extracted from images of maps. Identification of these high level drawing constructs requires considerable knowledge of the type of drawing concerned; it is rare for someone involved in manual drawing input to simply copy the document without giving at least some thought to its contents. Current commercial document input systems, however, only begin to exploit prior knowledge of drawing type and typically produce lower level representations than are really needed. Research results must be exploited, and further research is required, if useful interpretations are to be delivered at acceptable levels of speed and accuracy.

In writing this book we have sought to gather together relevant information on the current state of line drawing interpretation and present it, along with the underlying theory, in a manner accessible to both academics and to the actual and potential developers of industrial document interpretation systems. The aim of the book is to show how line drawing interpretation systems can be developed. To this end we consider the problems to be faced (Chapter1), overview the technology underlying line drawing interpretation (Chapter 2) and consider the key operations (Chapters 3-7). To help the reader to realise the technology, detailed descriptions are given of selected methods, algorithms and data structures. To provide examples of the formation of complete interpretation systems from the available components we consider our own and our close colleagues' work in some detail. The map (Chapters 8 and 9) and engineering drawing (Chapter 10) interpretation systems developed by Prof. Ablameyko's group at Minsk, Belarus are discussed, as is the ANON system developed by Drs Steve Joseph and Tony Pridmore in Sheffield, England. Systems developed by others and/or applied in other domains are discussed more briefly, not because they are any the less impressive or important, but to allow us to exploit our own experiences and produce, hopefully, a more accurate and immediate text. Line drawing interpretation is a challenging area with enormous practical potential; we hope that our book can convey something of the spirit of the endeavour.

Sergey Ablameyko	Minsk
Tony Pridmore	Nottingham

June 2000

Acknowledgements

We would like to express our immense gratitude to the Belarusian and British colleagues with whom we have worked on line drawing interpretation systems and techniques. In particular, our colleagues from the Laboratory of Image Processing and Recognition, Institute of Engineering Cybernetics, Minsk, Belarus and the Department of Mechanical and Process Engineering, University of Sheffield, Sheffield, UK

From Minsk, we particularly wish to thank V. Bereishik, O. Frantskevitch, M. Homenko, N. Paramonova, O. Patsko, V. Starovoitov, G. Aparin and D. Lagunovsky. Their support has allowed Prof. Ablameyko to develop several generations of line drawing image interpretation system over the last 15 years. Some sections of this book are drawn from papers previously written and published by Prof. Ablameyko in collaboration with these colleagues. Sections 5.6, 6.2.3, 6.3.3, 7.3.2, 7.6, 8.4, 9.1, 9.2, 9.4, and Chapter 10 benefit from contributions by V. Bereishik, O. Frantskevitch, N. Paramonova and M. Homenko. Dr. G. Aparin contributed to Section 8.6. The results presented in Section 9.3 were obtained in collaboration with Dr. V. Starovoitov. A. Kryuchkov developed the software described in Section 8.5. Finally, special thanks go to Prof. V. Tanaev for his permanent support of this research.

In the UK we wish to express our particular thanks to Dr. Steve Joseph. Dr Joseph first introduced Tony Pridmore to the line drawing interpretation problem and Dr Pridmore enjoyed many discussions on this and other subjects during his time as Research Associate in Dr Joseph's laboratory. The ANON system was developed as part of an SERC ACME project headed by Dr Joseph and the description of ANON given here draws heavily upon papers written and published jointly by Drs Joseph and Pridmore. Many thanks are due for his permission to use this material.

The initial idea for this book arose in the autumn of 1994, when Prof. Ablameyko visited Sheffield under a Royal Society Exchange Agreement. The Royal Society also supported the production of the text by awarding Tony Pridmore a Visiting Fellowship to the Institute of Engineering Cybernetics, Minsk for two weeks in the autumn of 1998. Without this support it is unlikely that the book would have been completed.

Equally important is the personal support provided by our families; particularly Natasha, Andrei and Masha Ablameyko, and Lynn, Ellen, Joe and Amy Pridmore. Many thanks to all of you.

Contents

1. The Line Drawing Interpretation Problem 1

   1.1 Motivation 1
   1.2 Manual Input vs Document Scanning..6
   1.3 Raster-Based vs Vector Representations 8
   1.4 The Interpretation Problem..9
   1.5 Engineering Drawings and Maps..10
   1.6 Line Drawing Interpretation, Image Understanding and Pattern Recognition.13
   1.7 Current Line Drawing Interpretation Systems...15
      1.7.1 Commercial Systems 15
      1.7.2 Laboratory Systems 16
   1.8 The Line Drawing Interpretation Literature 19

2. Components of a Line Drawing Interpretation System 21

   2.1 Design Criteria for Drawing Interpretation Systems 21
   2.2 Five Stages of Line Drawing Interpretation...24
   2.3 Intermediate and Target Representations 27
      2.3.1 Raster Representations 27
      2.3.2 Vector Representations..30
      2.3.3 Universal Drawing Entities..33
      2.3.4 Two-Dimensional Objects 36
      2.3.5 Three-Dimensional Shape and Semantics 37
   2.4 System Architectures and the Role of A Priori Knowledge 39
      2.4.1 Sequential Architectures..40
      2.4.2 Blackboard Architectures: The Expert System Approach 42

3. Document Image Acquisition 45

   3.1 Scanning Devices 45
   3.2 Image Coding 50
   3.3 Image File Formats...53

4. Binarisation 57

   4.1 A Taxonomy of Thresholding Techniques 57
   4.2 Document Image Statistics...61
   4.3 Binarising Line Drawings...64
      4.3.1. Point-Dependent Threshold Selection 64

    4.3.2 Region-Dependent Threshold Selection 71
    4.3.3 Image Partitioning 73

5. Binary Image Processing and the Raster to Vector Transformation 75

  5.1 Raster to Vector Conversion 75
  5.2 Some Definitions 77
  5.3 The Distance Transform 81
  5.4 Mathematical Morphology 90
  5.5 Reducing Noise in Binary Images 91
  5.6 Reducing Noise in Binary Images of Line Drawings 96

6. Analysis of Connected Components 101

  6.1 Nomenclature 101
  6.2 Contouring 102
    6.2.1 Goals 102
    6.2.2 Classification of Contouring Techniques 102
    6.2.3 A Contouring Algorithm 104
  6.3 Skeletonisation 107
    6.3.1 Motivation 107
    6.3.2 Approaches To Thinning 109
    6.3.3 A Thinning Algorithm 111
    6.3.4 The Medial Axis Transform 113
    6.3.5 Removing Noise from Thinned Data 115
  6.4 Grey Level Skeletonisation 117

7. Vectorisation 119

  7.1 Approaches To Vectorisation 119
  7.2 Global Vectorisation Methods 122
    7.2.1 Iterative Methods 122
    7.2.2 Feature-Based Techniques 124
    7.2.3 Hybrid Approaches 125
  7.3 Local Vectorisation Methods 127
    7.3.1 Line Following Methods 127
    7.3.2 Raster Scan Techniques 128
  7.4 The Hough Transform 130
  7.5 Direct Vectorisation 133
  7.6 A Vector Database 137
  7.7 Removing Noise from the Vector Model 140
  7.8 Alternative Raster to Vector Technologies 141

8. Interpreting Images of Maps 143

  8.1 Introduction 143
  8.2 System Overview 143
  8.3 Map Interpretation Principles 145
  8.4 A Classification of Map Entities 146
    8.4.1 Line Objects 146
    8.4.2 Area Objects 147

    8.4.3 Symbols 151
  8.5 Interactive Map Interpretation 152
    8.5.1 Automated Image Interpretation under Operator Control 152
    8.5.2 Operator-Supplied Context 157
  8.6 Output Formats 159
  8.7 Quality Issues 160
    8.7.1 Basic Concepts 160
    8.7.2 Maintaining Quality during Interpretation 161

9. Recognising Cartographic Objects 163

  9.1 Recognising Isolines 163
    9.1.1 Overview 163
    9.1.2 Theoretical Background 164
    9.1.3 An Isoline Recognition Procedure 165
  9.2 Recognising Roads 169
    9.2.1 Overview 169
    9.2.2 Interpreting the Road Layer 170
    9.2.3 Extracting Roads and Correlated Objects from the Black Layer 172
  9.3 Recognising Texture and Area Objects 176
    9.3.1 Overview 176
    9.3.2 Texture Border Definitions 176
  9.4. Recognising Symbols 180

10. Recovering Engineering Drawing Entities from Vector Data 185

  10.1 Design Principles and System Architecture 185
  10.2 Vectorisation and Entity Recognition Processes 188
  10.3 Extracting Arcs and Straight Lines 192
  10.4 Recognising Crosshatched Areas 195
  10.5 Recognising Dimensions 199
  10.6 Detecting Blocks 203

11. Knowledge-Directed Interpretation of Engineering Drawings 209

  11.1 An Image Understanding Approach 209
  11.2 Drawing Entities as Schemata 212
  11.3 Image Analysis Facilities 214
  11.4 The ANON Architecture 215
  11.5 Control Issues 217
  11.6 Entity Extraction: Chained Lines 219
  11.7 Top-Down and Bottom-Up Control 224
  11.8 Performance 226
  11.9 Scene Formation 233
    11.9.1 Searching Image and Memory 233
    11.9.2 Coincidence Links 236
    11.9.3 Partial Interpretations and Schema Fusion 239
  11.10 ANON in Context 240
  11.11 Discussion 241

12. Current Issues and Future Developments 243

12.1 Higher-Level and 3D Representations 243
12.1.1 Resolving Inconsistency 243
12.1.2 Editing and Parameterisation 244
12.1.3 3D Reconstruction 250
12.2 Exploiting Domain Knowledge 253
12.3 The Role of the Operator 256
12.4 Performance Measures 257
12.5 Topics for Future Development 258

References 263

Index 281

Chapter 1
The Line Drawing Interpretation Problem

1.1 Motivation

It is often said that a picture paints a thousand words. Unusually for such clichés, professionals in a wide range of areas and disciplines demonstrate the truth of this every day as they use drawings to record, communicate and test their ideas. From rough sketch maps to intricate blueprints, drawings are a ubiquitous and highly valuable tool.

Line drawings are particularly useful (figure 1.1). Our human visual system is extraordinarily good at making sense of even the crudest linework. The grossly exaggerated facial features typically used in newspaper cartoons, for example, rarely prevent their subjects from being recognised, even by readers unfamiliar with the particular cartoonist's style. In almost complete contrast, line drawings may also be used to record highly accurate quantitative measurements in the form of scale technical drawings and maps. Moreover, most people can create most types of line drawing, with simple equipment and only a moderate amount of training and practice.

As the price/performance ratio of computer technology continues to decrease, more and more of those who spend time working with drawings find that they also work with computers. Even the more basic of today's machines can support graphics-based packages that allow their users to both create and manipulate images and drawings. Lines may be moved, copied and resized. Their colour and width can be altered, as can their style: arrows, dashed, dotted and chained lines can be created at the press of a button. Groups of lines and regions may be dealt with in a similar manner. Printers capable of high quality rendering of the results are now within the budget of most home computer users. The full potential of the available technology cannot be realised, however, if the drawings with which one must work are stored on paper.

Engineering design provides an example. Sophisticated Computer-Aided Draughting or Design (CAD) tools are now readily available. Some of these can be interfaced directly to Computer-Aided Manufacturing (CAM) systems, forming integrated environments in which the entire product development process is supported by powerful software tools. CAD/CAM, sometimes known as Computer Integrated Manufacturing (CIM), is an attractive proposition and may already be applied to some engineering tasks.

a.

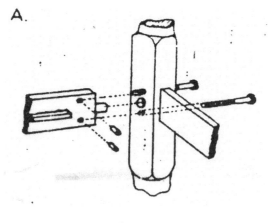

b.

The Line Drawing Interpretation Problem 3

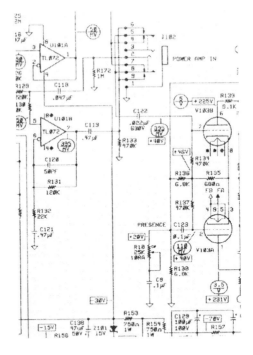

c.

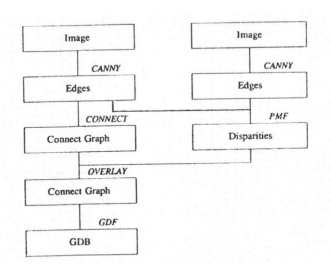

d.

Figure 1.1. A few of the types of line drawing commonly found in science and engineering, a) a perspective drawing of a natural scene, b) an annotated perspective drawing showing how several parts should be combined, c) an electrical schematic, d) a simple flow chart.

Many situations, however, require only comparatively small modifications to be made to existing designs. When the work is carried out within a CAD/CAM system, that system can be used to make and record the necessary modifications. If the original design only exists in paper form, a problem arises. A choice must be made between recreating the existing design information in an appropriate electronic form or continuing to work on paper.

Continued work with paper is immediately attractive, being both cheap and simple: merely photocopy the drawings concerned, whiting out the areas to be changed with a piece of blank paper, then draw in the new version. There is of course a secondary and longer-term cost associated with this approach; investment made in CAD/CAM systems does not bring the returns that it should. A situation persists in which some designs are on paper and some on CAD, necessitating different procedures for different projects. Despite both this and the widespread take-up of CAD technology, a return to paper is sufficiently tempting that stocks of paper-based engineering drawings are still growing faster than are libraries of CAD system files.

The availability of computer-based tools which can either perform or at least support the input of paper line drawings would benefit all of the engineering disciplines. Mechanical piece-part drawings might be input to CAD/CAM systems. Electrical schematics could be input to Electrical CAD systems, circuit simulation packages and/or knowledge-based systems seeking to identify faults in a circuit or design. Architectural drawings might be input to systems that create virtual reality walk-throughs of the buildings they describe. Computer-Aided Software Engineering (CASE) tools are becoming much more common. The design methodologies they support typically rely heavily on, for example, data flow or entity-relationship diagrams and structure charts. These are all types of line drawing which often exist on paper and might usefully be input to software tools that can manipulate and modify them.

A similar situation arises from the widespread use of maps. A huge number of maps of various different types exist. These must often be modified, new ones produced, and a range of tasks performed which rely on or otherwise involve the use of some kind of map. Manual map production and modification is a very slow and expensive process. At present, so-called Geographic Information Systems (GISs) are being applied in all areas of map generation and processing. GISs do not employ paper maps. Instead, they provide easy to use, powerful and flexible interfaces through which their users can manipulate various electronic representations of the relevant information. A GIS user can perform many map-related tasks much more quickly and efficiently than would be possible using only paper-based representations. If the benefits of this technology are to be obtained, however, the necessary maps must be input into a GIS and presented in one of the required formats. This is potentially a huge task. It is made larger by the number of different map types involved and the differing requirements of application areas as diverse as agriculture, ecology, geology, forestry and the military. The obvious benefits to be accrued from automatic map interpretation have attracted a significant amount of both industrial attention and academic research effort.

Although we shall focus in this book on the problems of, and techniques employed in, the interpretation of engineering drawings and maps, there is a clear requirement for fast, precise and inexpensive input of paper-based line drawings into computer systems in a diverse set of application domains.

The problem of text recognition in particular has attracted much attention. Handwritten text remains difficult to deal with. In some circumstances, however, text recognition systems can provide fast input of printed text, saving much human time and effort. A well-developed ability to input textual documents would provide a significant boost to office automation and ease the transmission of information across computer networks. Many office computers are now fitted with fax modems, which can send and receive images of computer screens. The use of the facsimile standard means that images can be generated and received all around the world. Unfortunately, documents stored in fax format cannot be modified by word processing or desktop publishing software. Systems capable of reliably creating, for example, ASCII text files from fax images would be of great value here. Further motivation for work in automatic text recognition comes from interest in the creation of digital libraries storing large numbers of documents.

In general, documents may contain graphics and photographs as well as text. Automatic line drawing input, along with the ability to recognise, for example, that some part of a page contains a photograph, would clearly be of value in each of the above situations. This is a specific example of the more general problem of automatically characterising page layout; identifying tables, text blocks, footnotes, graphics etc.

Research into Computer-Supported Co-operative Working (CSCW) environments has emphasised the role played by sketches and diagrams in collaborative work. The classic co-operative work scenario involves a number of colleagues standing around a blackboard upon which a solution to some problem is being developed via a process of debate and negotiation. Line drawings may be used either during the development of the solution or to (partially) express the final result of the collaboration. CSCW projects aim to provide hardware and software which makes this type of working environment available at a distance, across computer networks. Instead of being physically close, those involved would each run software allowing them to access and update in real time a common database which records the developing solution. CSCW environments would be greatly enhanced if their users could input existing paper documents and drawings, perhaps referring to similar problems and solutions, directly into the evolving database.

Much was made during the 1980s of the notion of the "paperless office". While many of the problems and scenarios outlined above clearly contribute to this rather ambitious aim, the focus of current work in document and drawing analysis is somewhat different. It is now recognised that paper has a continuing role to play; many people prefer to read, annotate and disseminate hard copies of documents, even if they have to print them off from their computers first. The current aim of workers in drawing and document image analysis is not to replace paper, but to integrate paper and electronic documents as efficiently as possible. The availability of effective document image analysis systems would mean that a medium would exist which could be read by computers and people alike.

A detailed review of all the potential applications of automatic line drawing input is beyond the scope of the current book. The short analysis given here, however, is enough to indicate that most computer-based systems working with graphical information stand to benefit from the development of techniques and devices which support fast input of the information contained in paper-based line drawings. For the present we shall concentrate on our prime areas of interest: the machine interpretation of engineering drawings and maps.

1.2 Manual Input vs Document Scanning

Special tools and devices for digitisation of graphical information have been available since the early days of computing. The Graphical Information Systems created in the 1970s and developed in the early 1980s were based mainly on manual input of graphics. The digitisers they used were at best semi-automatic and involved the operator tracing over lines, curves and objects on the document with some type of pointing device. The position of this pointer was read by the machine and strings of pointer locations used to form a representation of the underlying drawing. Text and other semantic information were also input manually, usually via a keyboard. This process suffers a number of drawbacks:

- It is very slow. Input of an average A2-A3 size map, for example, may take tens of hours. Large and complex drawings may require hundreds of hours of effort.

- It is prone to error. Accurate manual input of object co-ordinates is especially difficult. Many documents to be traced are accurate to fractions of a millimetre; the computer systems to which they are to be input may represent images and drawings to a resolution of 300 dots per inch (12 dots per mm). It is practically impossible to extract co-ordinates manually to this level of precision.

- It is very hard work. The operators of manual digitisation systems usually become tired very quickly. As a result the quality of the digitisation produced varies considerably over time. In the longer term this type of work can adversely affect the operator's vision.

The difficulties listed above mean that companies taking this approach often require many operators and digitisation systems. The financial costs of both setting up and maintaining a manual digitisation capability are therefore high. A number of companies in the UK still perform manual drawing input on a bureau basis, the typical cost of inputting an A0 drawing to a standard CAD system, for example, is £200-300. In many cases the expenditure of such a large amount of effort and/or money may simply not be worthwhile. If engineering drawings are being input to CAD to allow modification of a design, that design might have to be updated many times before the investment involved is recovered. In a reasonably sized project several tens or even hundreds of drawings might have to be input.

In the mid to late 1970s, document scanners began to appear and be used, primarily to input text-based documents. The main benefit of scanners is that they allow fast, automatic transformation of an image from analogue (i.e. paper-based) to digital form. The result of scanning a document is a digital raster file. This contains a digital image: a two-dimensional matrix of known dimensions in which each matrix element or pixel (picture element) contains a (usually integer) value proportional to

The Line Drawing Interpretation Problem

the brightness of the corresponding area of the scanned document. This value is often referred to as a grey level. Figure 1.2 shows an example. Good quality document scanners are now available from High Street computer stores for a few hundred pounds. They provide a significant reduction in input time, compared to manual methods, but produce a lower-level representation of the input drawing.

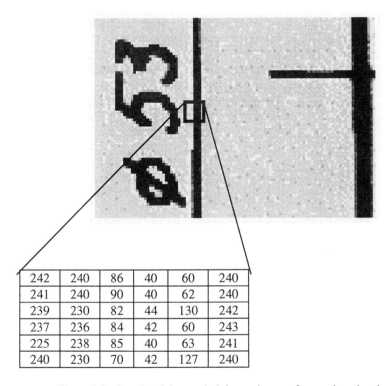

Figure 1.2. Grey level data underlying an image of an engineering drawing.

During manual input the operator typically uses his/her knowledge of the type of drawing concerned to recognise and label, perhaps via a keyboard, important drawing features. It is rare for someone involved in manual drawing input to simply copy the document without giving at least some thought to its contents. In contrast, scanners merely photograph drawings, performing no recognition or interpretation whatsoever. As a result, raster images are not usually employed directly in graphic information systems except for low-level functions such as information compression, storage, transmission and display. It should be noted, however, that raster images remain a commonly used drawing representation format.

1.3 Raster-Based vs Vector Representations

Any automatic drawing input facility should obviously produce representations that support the tasks that need to be performed with or on the input drawing. At the highest level each type of drawing will have its own, application-specific target representations constructed from sets of symbols denoting particular features of the input document. Given a cadastral map, for example, the goal may be to identify rows of houses. There are, however, intermediate representations that may be extracted from a wide variety of drawing types and which are of use in a wide range of application areas.

Many drawing manipulation systems, including most CAD and GIS systems, have a well-developed ability to process graphics represented in so-called vector form. In a vector representation sets or strings of simple curves describe each primitive or object. Straight segments form the basis of most vector representations, hence the name, though circular arcs are quite common and some schemes also incorporate ellipses.

Vector-based representations are widely used: raster images and vectors are now the two main formats for line drawing storage. Raster images have a number of advantages over vector-based methods of data representation. These are:

- a simple, generic data structure, i.e. a matrix of pixels;

- easy, direct visualisation on a variety of image display devices;

- the possibility of direct and fast draughting on raster plotters;

- increased accuracy over manual digitisation methods.

On the other hand, direct use of raster data for drawing manipulation suffers from a number of drawbacks, which make it difficult to create commercially applicable systems. These are:

- the large volume of raster data generated by high resolution scanning;

- the absence of logical object structure in the raster data. Object editing in particular is very difficult; to remove, for example, a centre line from an engineering drawing one must modify all the pixels contributing to that line. It is much easier to extract objects from vector than from raster-based representations.

Scanners and the associated raster-based document processing technology can free humans from the difficult task of manually inputting line drawings, speeding up the process and avoiding the many errors that typically result from manual digitisation. Raster-based techniques increase the level of automation of the graphics input and design processes and can aid the execution of many tasks in agriculture, industry and other areas relying on digital maps and/or drawings. The representations produced, however, leave much to be desired. If the full potential of the technology is to be realised, scanned raster data must be converted into a symbolic form which is more suited to subsequent drawing manipulation tasks. Although the desired representation may be both high-level and application-specific, vector-based descriptions have a large part to play.

1.4 The Interpretation Problem

The key question facing those concerned with the input of line drawings to computers is how appropriate symbolic representations can be extracted from the image(s) provided by a typical document scanner. Any system that can interpret a line drawing image to produce such a representation will almost certainly comprise a large software component, though it may also incorporate some special purpose hardware. An ideal system would be fully automatic. A semi-automatic process combined with a (preferably) high-level user interface would, however, be acceptable in most domains.

There now exist graphical input systems, including some commercial products, which extract vectors from digital images. A peculiarity of existing systems is that, despite producing only low level representations, they are usually oriented to one type of drawing. Few vectorisation systems are truly generally applicable. Many current map interpretation systems, for example, deal only with cadastral maps. It is evident that successful interpretation of any line drawing requires considerable knowledge of the specific features of drawings of that type. There are, however, great benefits to be obtained from the digitisation of all types of drawing. As a result, the efficient representation and use of appropriate available knowledge is a key issue in line drawing interpretation.

Many objects depicted in line drawings are either explicitly (graphically) or implicitly (by agreed semantics) connected to each other, forming scenes. Prior knowledge of the scene structure of, and likely object relations within, a drawing eases the task of extracting objects from the image. Fewer object classes need be considered at any given time, speeding up the interpretation process. Hypothesise and test strategies may be used to determine object class and to identify other semantic characteristics. If this is to be achieved, however, knowledge of possible relations between both individual, concrete objects and between object classes is required.

Part of this knowledge is naturally embedded in low-level object extraction software (for example, "a circle is a set of points having equal distance from a centre", "crosshatching is a closed area filled by thin parallel lines which have approximately equal distance from each other", etc.). More can be applied after a vectorisation stage (by noting, for example, the maximal width of thin lines and the minimal width of thick lines). The majority of the a priori information, however, should be thought of as expert knowledge and explicitly recognised as such, perhaps by its being stored in a distinct knowledge base. This idea is developed further in Chapters 2 and 8-11.

Line drawings, particularly engineering drawings and maps (and hence their images), are often very large, taking significant amounts of memory to store and significant amounts of time to process. Scanned images vary in size from 2550 x

3300 pixels for a business letter digitised at 300 DPI (dots per inch) to 34000 x 44000 pixels for a 34" x 44" E-sized engineering diagram digitised at 1000 DPI. At 200 DPI an E-Size drawing generates 8 megabytes (Mb) of raw data. At 400 DPI this becomes 32 Mb. At 1000 DPI, the minimum acceptable for graphics arts applications, we require 25 * 8 or 200 Mb. At 2000 DPI, the ideal graphics arts resolution, 800 Mb are needed. As is to be expected, there is a trade-off of resolution against scanning time; high resolution images take longer to acquire but contain more information than lower resolution ones. Most current drawing interpretation systems rely on powerful workstations or specialised hardware designed to process large volumes of data. These machines can be expensive and may therefore only be available to larger companies. Drawings are, however, used in all kinds of enterprise. Hence there is a market need for systems which can run on more commonly available, personal computers. Such systems are under development.

1.5 Engineering Drawings and Maps

In general, images of line drawings have a number of properties that distinguish them from other images of, for example, real world scenes containing people or landscapes. First, they comprise mainly dark lines on lighter backgrounds. The typical line drawing image is composed of an often large number of elongated regions of dark pixels, each corresponding to one or more pen strokes on the original drawing. In all but the best quality drawings there will be variations in grey level (pixel value) within these regions. These are due to variations in pen pressure, irregular ink flow, imperfect pen tips or imperfect scanning. The paper background, though generally much lighter than the ink marks, is also unlikely to be uniform. Finger marks, ink smudges, variations in the texture of the paper and, again, imperfect scanning will introduce sometimes significant variations into the grey levels of the image regions corresponding to blank paper. Most of the boundaries between inked and plain paper will be well defined. In some circumstances, however, the contrast of this boundary may not be high. Draughtsmen tend, for example, to begin to lift the pen from the paper as they near the ends of lines. This can mean that some lines gradually fade away, their grey levels merging into those of the background paper.

Each drawing type also has its own particular features. As prior knowledge of the class of drawing to be input seems likely to figure in any successful interpretation system, we now consider specific properties of the documents with which we are most concerned; engineering drawings and maps. These types of line drawing have the following distinctive features:

- Mixed text and graphics. In text-based documents graphics and text are usually placed in separate areas; in maps and engineering drawings text, symbols and graphic objects are closely mixed and may even overlap.

- Multiple object types (different kinds of lines, symbols and regions).

The Line Drawing Interpretation Problem 11

- Objects of any given type may occur anywhere in the drawing, in any orientation and at any scale. Text may appear hand-written or printed in arbitrary fonts.

- The drawing could be coloured, and its quality is likely to be far from ideal.

Let us now consider maps in a little more detail. Automatic interpretation of several map types has been attempted. These include cadastral maps, city maps, large scale maps with scales from 1:500 to 1:10 000, separate map layers and coloured topographic maps with scales from 1:10 000 to 1:200 000 and geographical survey maps with scales of 1:500 000 and above. Many systems have been developed and used to input information automatically from large scale maps (1:500 to 1:10 000) which have low linework density and little variety of information [1]. Interpretation of medium scale maps (1:10 000 to 1:50 000), which have a higher linework density and convey a much greater variety of information is typically limited to the acquisition of single layers of information, for example, road nets.

Cadastral maps are the focus of several current automatic map interpretation systems. These have approximately the same view and object parameters (thickness of parcel borders, orientation of hatched lines to represent houses or administrative buildings, localisation of text, etc.) regardless of their country of origin (figure 1.3). There is also a direct correspondence between objects in the image and objects in the world they represent, which can be exploited during the interpretation process.

Figure 1.3. A section of a grey level image of a cadastral map.

Topographic maps provide another important class of map images. These may be coloured or black-and-white (figure 1.4). Each country has its own standards for map production but in general every map contains three primary object types:

- lines, which must be represented in the output database by their medial axes;

- symbols, which have restricted geometrical parameters and must be represented by one or two points;

- regions, which must be represented by their bounding contours.

Map overlays also need to be input to computer systems. The overlay is a special form of geographic map in which each layer displays only one of the map's several colours (figure 1.4). Each layer is a black-and-white sheet that depicts objects of one colour on the original map. The basic layers are: roads (yellow on the original map), hydrography (blue), forestry (green), isolines (brown) and a default layer containing black objects. A complete digital map is obtained by processing each layer separately, then combining the results.

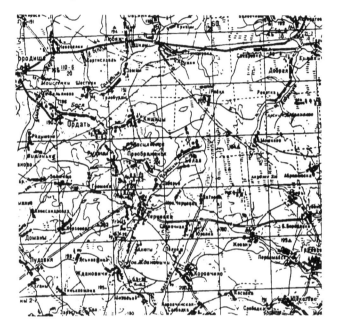

Figure 1.4. A grey level image of a section of topographic map.

Maps and engineering drawings are similar in that both effectively provide an annotated image of the object(s) of interest. Engineering drawings usually comprise lines in a range of widths along with a variety of associated symbols and text. Some of the lines (usually the thicker ones) represent the projection onto a plane of the contours of an object or object section. The remainder of the drawing consists of thinner lines, symbols and text. This second component is much more linguistic or symbolic than the first, pictorial component. It is generally read, rather than perceived, and conveys any additional information needed to support understanding of the geometric sections of the document. A comparatively simple engineering drawing is shown in figure 1.5.

The Line Drawing Interpretation Problem   13

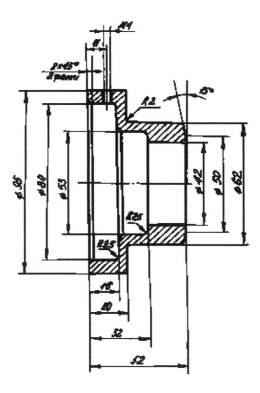

Figure 1.5. An image of a mechanical piece-part drawing

1.6 Line Drawing Interpretation, Image Understanding and Pattern Recognition

Line drawing interpretation may be thought of as a subset of document image analysis, which in turn is a subset of the much wider field of image understanding and computer vision.

Given one or more input images, vision aims to describe the real-world situation which generated those images, i.e. to make explicit the information about the world that is only implicit in the pixel arrays. Workers in computer vision and image understanding seek to understand the processes and representations underlying vision in sufficient detail that they may be implemented on a computer. This effort may be motivated by a desire to understand how, for example, human vision might work, or a wish to create practical computer vision/image interpretation systems.

It is important at this point to make clear the distinction between image understanding and image processing. Image processing is perhaps the more familiar and involves the modification of one image to produce another. The area is well developed and many very useful techniques now exist. All these processes, however, map images onto images. The image produced is usually for human consumption; the image processing operations applied typically seek to make some aspect or feature of the input image more easily visible to the human viewer. Image processing does not, however, attempt to extract the information within the image, only to make it more accessible. While many image understanding systems employ image processing methods, the goals of image processing and computer vision systems are very different.

Vision is a complex task. Each pixel value is a complex function of the reflectance and geometry of the viewed surface, its illumination and the relative position and orientation of the observer. Any successful vision system, biological or artificial, must invert this complex and so far unknown function. Vision is clearly an underdetermined problem; the image alone does not provide all the information required for its solution; some a priori knowledge or assumptions are necessary. Vision problems and systems may be broadly classified according to the amount of a priori knowledge available to them.

Given an appropriate description of the exact 3D shape of an object and a grey-level image of that object, model-based vision systems must determine the 3D position and orientation of the object relative to the viewer. To achieve this, an object recognition phase is required during which the system must match the image to the available models to determine which, if any, are visible [2]. Model-based vision has a wide variety of applications in industry, and a number of model-based systems are quite readily available. The model-based approach is fine if precise models of the geometry of all possible objects of interest are available. Modelling all such objects to high precision is, however, a tall order.

Knowledge-based vision systems operate without quantitative geometric models. Given an image and some knowledge of the type of object to be expected, a knowledge-based vision system must locate objects in the image, i.e. segment the image or data extracted from the image into a number of regions (one region per viewed object), then describe each object to the best of its ability.

Both model- and knowledge-based vision systems typically rely on features extracted from the image. These features are functions of grey level, which itself is a complex function of surface reflectance, geometry, illumination and viewpoint. The third class of vision system attempts to separate these effects and extract features of the viewed world. Such features provide much better input to model- and/or knowledge-based systems. No hard a priori knowledge is available. The goal of these "pure" vision or "shape-from-X" systems is to compute features like distance, orientation, etc. from images of any scene. If this problem is to become tractable assumptions must be made. A major research question in shape-from-X methods is what assumptions are both useful and near enough true. There exists a range of shape-from methods; shape from shading, shape from motion, shape from texture, etc.

It is apparent from the above that line drawing interpretation is an example of a knowledge-based vision problem. No reliable, fixed and quantitative models of CAD entities and cartographic objects are available, so a true model-based approach is not possible. Though many workers in line drawing interpretation write of "recognising" drawing objects, it should be stressed that this is not recognition in the sense which is commonly accepted in the wider computer vision community. Line drawing images, however, are really quite highly constrained. The goal of a map or engineering drawing system, therefore, is to segment the image into regions corresponding to drawing entities and provide as much information as possible about each such entity.

Line drawing interpretation is sometimes thought of as a problem in pattern recognition. The general pattern recognition task [3] is to determine to which of a predefined set of classes a given pattern belongs. Although quite complex descriptions of the input data may be extracted for use within the classification process, the sole result of pattern recognition is a symbolic label denoting the class of the input. In contrast to this "classical" pattern recognition task, to "recognise" a line-drawing object means to extract that object from a line-drawing, establish its class, compute its position within the drawing and represent that and any other relevant geometric and/or topological parameters of the object in a pre-specified format. These extended requirements make line-drawing object recognition more complex than classical pattern recognition.

Line drawing interpretation is a rich and challenging problem; research in this area draws on results in image processing, pattern recognition, computer vision and knowledge-based systems. Any successful solution to the drawing interpretation problem will comprise elements of all these disciplines integrated and fused together by knowledge and experience of the task at hand. Considerable progress has already been made.

1.7 Current Line Drawing Interpretation Systems

In the mid to late 1980s, document image analysis began to grow rapidly as results achieved in hardware development allowed images to be processed at acceptable speeds and reasonable cost. We now briefly review existing line drawing interpretation systems.

1.7.1 Commercial Systems

Document processing systems capable of storing forms and other structured documents, performing Optical Character Recognition on typewritten text and compressing drawings are now commercially available. There are many systems on the market which perform or support vectorisation of line drawings. It is not possible to describe all these products, but we can acknowledge systems developed by the following companies: Able Software Company, AccuSoft Corp., Alpharel, Arbor Image Corp., Audre, ColorSoft Inc., CPI, Digist Software, ERDAS Inc., Grumman InfoConversion, GTX Corporation, Horizons Technology, InfoGraphix Technologies, Intergraph Corp., MicroImages Inc., MST, Pacific Gold Coast, Rorke Data Inc., Softdesk Imaging Group, Sovereign C.S. Ltd., Vidar Systems Corp. [4].

Many vectorisation systems are currently available on the Russian and Eastern European markets. In the USA and Western Europe the primary development of GIS technologies took place at a time when scanners were very expensive and hence rarely used. As a result, most cartographic data was input to the new systems via manual digitisers. In Russia, interest in GISs has developed only recently, after scanners and scanning technology had become inexpensive and readily available. Document image processing and vectorisation systems were therefore developed early and transferred to personal computers very quickly. Moreover, the rapid growth of GIS systems in the East has made scanning technologies very popular there, with scanners being used in practically all map digitisation.

To give a clearer picture of the facilities offered by current vectorisation systems, we consider just one of these products, Visus [5]. Visus takes as its input a binary image; acceptable input formats are PCX, B/W TIFF, Vidar Compressed, CAD Camera, SGI, PROCAD and uncompressed bitstream. The system is menu or command line driven, at the user's choice. A number of image manipulation and noise removal functions are also incorporated into the system, providing a significant preprocessing facility. Visus is capable of extracting lines, polylines, arcs, circles, fill, line ends and intersections, outlines, lines contributing to arcs and circles, dots, elliptical dots, squares, and rectangles. Output may be expressed in a variety of formats, including AUTOCAD DXF, IGES, HPGL, HIPLOT, GERBER ASCII, Documented Binary, DXB and Calcomp 960. The time taken to vectorise a raster image obviously depends upon the options selected. A 200 DPI drawing 36" X 48" may, however, be converted in less than 15 minutes. An A4 drawing takes about five minutes on a medium specification PC.

1.7.2 Laboratory Systems

Visus was developed some ten years ago but remains typical of commercial interpretation systems. There exists a number of more advanced systems, operational in research laboratories but not yet developed into commercial products. We now briefly consider just some of them. Our review is limited to engineering drawing and map interpretation systems and it must be stressed that what follows is not a complete overview, even of those areas. Our aim is simply to give an impression of the range of approaches being considered. Further details of some of these systems and the techniques they employ are given in later chapters. We begin with research into engineering drawing interpretation.

A multi-component system interpreting mechanical engineering drawings is described by Kasturi et al [6]. This includes algorithms for text/graphics separation, recognition of arrowheads, tails, and witness lines; association of feature control frames and dimensioning text with the corresponding dimensioning lines; detection of dashed lines, sectioning lines and other objects. A modified version of the system has been used to interpret telephone system manhole drawings [7].

The system described in [8] identifies object outlines and dimension sets and combines them to produce 3D wire-frame models from images of paper-based engineering drawings. The system has been successfully applied to drawings of objects composed of planar, spherical, and cylindrical surfaces. A knowledge-based approach is taken, employing evidential reasoning and a wide range of rules and heuristics.

MDUS [9-11] was developed to convert mechanical engineering drawings into the Initial Graphic Exchange Specification (IGES), an accepted standard for exchange of graphic information among CAD/CAM systems [12]. The system is based on three levels of drawing interpretation: early vision (the lexical phase), intermediate vision (the syntactic phase) and high-level vision (the semantic phase).

MDUS supports identification of bars (straight segments), circular arcs, arrowheads and textboxes (regions in which text is expected to appear).

A system for interpretation of engineering drawing images is described in [13]. A vectorisation process is used which is similar to one proposed for use with map images, although there are some small differences. An object extraction stage then aims to produce a representation in terms of universal entities: arcs, circles, line types, blocks, crosshatched areas, dimensions, etc. Specific features of engineering drawing images are expressed in a knowledge base and widely used at both vectorisation and object location stages.

ANON [14] is based on the combination of schemata (or frames) describing prototypical drawing constructs with a library of low-level image analysis routines and a set of explicit control rules. The system works directly on a raster image without prior vectorisation, combining the extraction of primitives with their interpretation. The system integrates bottom-up and top-down interpretation strategies into a single framework.

CELESSTIN [15] relies on high-level knowledge to interpret images of mechanical drawings. Technologically significant entities are extracted and the whole set-up is analysed using kinematic knowledge of the drawn objects. The latest version of CELESSTIN uses knowledge of the functionality of extracted objects, rather than being limited to knowledge of object shape.

The REDRAW system was developed to interpret mechanical drawings and maps [16]. It uses a priori knowledge to achieve interpretation at a semantic level and is based on a model-driven system that can be completely parameterised. A priori knowledge of the domain induces a different interpretation process for each document class. Objects identified by REDRAW include, among others, parallel lines and hatched areas.

MARIS [17] has been used to digitise large, reduced-scale maps of Japan. A vector database is created and forms the basis of object recognition. Interactive editing and correction of mis-recognised and unrecognised objects refines the automatically produced results. MARIS supports the digitisation of three map layers: building lines, contour lines and lines for railways, roads and water areas.

PROMAP was developed to digitise German topographic colour maps with a scale of 1:25 000 [18]. The maps are scanned with a 24-bit RGB (red, green, blue) scanner and separated into four layers. Symbols and objects are identified at the raster level and further vectorisation is performed. A particular strength of PROMAP is its use of large amounts of a priori knowledge.

A system that automatically extracts information from paper-based maps and answers queries related to spatial features and the structure of the geographic data they contain is described in [6,19]. Algorithms to detect symbols, identify various types of line and closed contour, compute distances and find shortest paths have been realised within this system. Its query processor analyses queries presented by the user in a predefined syntax, controls the operation of the image processing algorithms, and interacts with the user.

Boatto [20] describes a system capable of automatic input and interpretation of Italian land register maps. This system makes extensive use of the semantics of land register maps to obtain correct interpretations. The final graph-based representation provides a formal description of topological and metrical properties that are stored in a database. The system requires the operator to resolve ambiguities and correct errors in the automatic processes.

A system that digitises colour and black-and-white Russian topographic maps with scales from 1:25 000 to 1:200 000 is described in [21,22]. The system combines automatic vectorisation and interpretation with interactive tools that support editing and input of unidentified objects. Vectorisation techniques have been developed which can process large images using restricted computer memory (the system is based on an IBM PC). An intermediate vector representation is built based on the notion of segments bounded by end and node points. Isolines, roads and region-based layers are extracted automatically. A combination of automatic and interactive techniques allows complex maps to be digitised in reasonable time scales.

A system capable of the robust identification of drawings superimposed on maps is described in [23]. First, graphic and character regions are decomposed into primitive lines, the definition of which includes their orientation and connections to other lines. Objects are extracted from these primitive lines by recognition techniques that use shape and topological information to group them into meaningful sets such as character strings, symbols, and figure lines. The system has been used to interpret equipment diagrams superimposed on maps.

An experimental parallel system, interpreting images of road maps and based on a multi-layer partitioned blackboard model, is described in [24]. The system was realised on the parallel AP1000 computer produced by Fujitsu. The authors were not, however, satisfied by either the recognition ratio or processing speed of their system.

Ah-Soon and Tombre [25] present a system which constructs models of rooms from images of architectural plans. Low-level features such as lines, arcs and loops are first extracted from the input drawing. Two complementary high-level interpretation techniques are discussed: one based on geometric analysis and symbol recognition, the other on spatial analysis.

An interesting approach to on-line interpretation of sketched engineering drawings is proposed in [26]. The Designer's Apprentice is a pen-based system that supports the production of detailed engineering drawings on a realistic electronic drawing board. The system makes an early distinction between object lines and associated annotation, which are then processed separately. Annotations are recognised using expert knowledge.

While work on improved low-level techniques continues, research groups working in both map and engineering drawing interpretation generally seek to raise the level of the representations produced. Engineering drawing entities, cartographic objects and scene descriptions are preferred to lower-level vector-based representations. The representation and use of knowledge is a key issue, as is the level and nature of the interaction required with the operator.

1.8 The Line Drawing Interpretation Literature

As line drawing interpretation systems and techniques have developed over the years many research papers have been produced. Several excellent bibliographies and edited collections of those papers now exist [27-30] and provide a valuable route into the wider literature. Some sections of the image understanding and computer vision literature are also relevant. Good introductions to and overviews of this area may be found in [31-35].

Papers on the interpretation of document or line drawing images now appear regularly in the image processing, image analysis and computer vision literature. Since the early 1990s, issues of several international journals have been devoted to line drawing interpretation. Special issues of the Computer Magazine (July 1992), Machine Vision and Applications (1992, 1993) and the International Journal of Pattern Recognition and Artificial Intelligence (1994) describe both systems and techniques for line drawing interpretation. A new journal devoted to the field, the International Journal on Document Analysis and Recognition appeared for the first time in 1998. There are now regular conferences and workshops in this area; descriptions of many drawing interpretation systems and methods can be found in the proceedings of the:

- International Conference on Document Analysis and Recognition (1991, 1993, 1995, 1997, 1999)
- International Workshops on Machine Vision Applications (1988, 1990, 1992, 1994, 1996, 1998, 2000)
- International Workshops on Document Analysis Systems (1994, 1996)
- International Workshops on Graphics Recognition (1995, 1997, 1999)
- Symposiums on Document Analysis and Information Retrieval (1992, 1994, 1996)
- SPIE Conferences on Document Recognition (1992, 1994, 1996, 1998)
- International Workshops on Advances in Structural and Syntactic Pattern Recognition (1990, 1992, 1994, 1996, 1998).

Though line drawing interpretation techniques are still developing and the problem is far from solved, the field has now reached a certain maturity. There is significant commonality to be found in the tools, techniques and system architectures employed across the area. Methods aimed at the interpretation of one drawing type are being more widely employed and methodologies for the evaluation of drawing interpretation technologies are being developed; both are sure signs of a maturing discipline. Although we cannot discuss every method, system and approach taken to line drawing interpretation we aim in what follows to give a feeling for the area, supported by more thorough discussion of selected techniques. In chapter 2 we overview the technology underlying line drawing interpretation before going on to consider the key operations in chapters 3-7. In chapters 8-11 we examine our own and our close colleagues' engineering drawing and map interpretation systems in some detail. We close in Chapter 12 with a discussion of possible future developments.

Chapter 2
Components of a Line Drawing Interpretation System

2.1 Design Criteria for Drawing Interpretation Systems

To date, image interpretation systems have been most successfully employed in industrial inspection tasks, bin picking exercises and for robotic guidance [36], though more recently attention has shifted towards more human-oriented applications such as face recognition and automatic visual surveillance. Most systems operate within structured environments where the data to be assessed is both well-known and minimised; the appearance of objects of interest can often be quite accurately predicted, with each object typically occupying only a small proportion of the image. Line drawings are a means of communication between humans. As they have evolved, drawing styles have emerged which rely heavily on the interpretive power of the human visual system. Hence, although line drawing interpretation might seem more tractable than a 'natural' vision problem like, say, the extraction of three-dimensional descriptions of walking people from a video tape of a street scene [37], the task is in fact highly complex.

At present, line drawing interpretation cannot be performed reliably without human intervention. The primary aim of research into drawing interpretation systems is to minimise that intervention, ideally, though not necessarily, producing a fully automatic system. With this background in mind we introduce the following set of design criteria for line drawing interpretation systems:

- The output of a line drawing interpretation system should be in a format that is suitable for direct input into the target application. The details of this requirement obviously vary from one form of drawing to another; the key here is that it should be possible to input the representations generated by drawing interpretation into the target system(s) without further work.

- Assuming that a fully automatic solution is not practical, drawing interpretation systems should comprise two clearly defined and separated parts. At present this usually means that the first should provide automatic image pre-processing and vectorisation and the second a combined automatic/interactive object recognition and interpretation system.

- Any vector representation(s) produced should describe all the significant pen strokes; loss of information at this stage could lead to incorrect object recognition and may restrict interactive processing. The operator must have access to the information required to make correct and consistent decisions.

- Any automatic image processing and/or vectorisation stages should produce the highest quality results possible. Vectorisation errors are likely to seriously disrupt both automatic and, to a lesser extent, interactive object recognition and will invariably increase the amount of interactive interpretation required. Inaccurate and misleading vectorisation could cause all the advantages of automatic interpretation to be lost.

- During object recognition a compromise must be made between automatic and interactive techniques to provide a balance between processing speed and accuracy of interpretation. Decisions as to which objects or parts of objects are to be extracted automatically and which interactively should be made as early as possible in the design process.

- The object recognition component of the system should allow the operator to correct interpretation errors with the minimum of effort. Any interactive object interpretation software must be carefully designed to be user-friendly.

- It should be possible to interrupt interpretation at any point while retaining the results of all processing performed thus far. This facility would support a human operator in monitoring the interpretation process. Should a particular operation, especially an automatic one, be seen to be introducing errors and/or distortions, the operator must be able to gain access to the results of previous processing stages.

- Constraints on processing speed obviously vary from application to application. As a general guideline, however, the time taken to input a document via the drawing interpretation system should be at most half that required to input the same drawing manually. This criterion broadly takes into account the financial costs incurred when replacing an existing manual system and reducing and/or retraining staff.

- The final representation should contain the maximum possible information about the input drawing while at the same time occupying the minimum possible data volume. There is clearly a trade-off here, the solution to which will depend upon the information content of the drawing and the environment and application for which drawing interpretation is required.

- The final representation must record the geometry of each drawing object to the required level of accuracy. Perhaps more importantly, the representation should, where possible, include information regarding the accuracy that can reasonably be expected given the techniques employed. This is very important, for example, in map digitisation as deviations in map object co-ordinates can have a significant influence on the value and future use of the resulting digital map.

- The final system should, as far as possible, require only standard (and therefore hopefully cheap) computer hardware. A solution based on PC technologies would be ideal. To rely upon expensive, special-purpose computing engines would be to limit the uptake of drawing interpretation technology among the smaller companies who could perhaps benefit the most from access to (semi-) automatic line drawing interpretation systems.

The design of a line drawing interpretation system is a non-trivial task. A clear understanding is needed of the likely content and structure of the input drawings as well as some idea of their expected quality. Detailed examination of a variety of target drawings must be an early step in any drawing interpretation project. A clear specification must also be drawn up, again as soon as possible, of the output required of the system.

Once the problem has been carefully considered and understood as far as possible, the next step is usually to consider which of a comparatively small number of commonly used processing stages are most appropriate to its initial solution. These standard processes and the representations they employ are the building blocks from which current line drawing interpretation systems are constructed. It should be stressed, however, that the identification of which stages are appropriate is by no means the final step in the process. Each may be implemented and used in a variety of ways; detailed consideration of the required output, the input drawing and the available a priori knowledge is typically required before detailed design decisions can be made. A suitable system architecture must also be designed and the proposed system implemented and tested. Moreover, when seeking an interpretation methodology for a new set of drawings it is often found that significant development work is required to modify existing techniques and processes to suit the task at hand. The same is true of system architectures; while a number of basic designs are in common use (see Section 2.4 below), modifications and developments are often necessary to bend a given architecture to a new application area.

In what follows we outline the basic components of current line drawing interpretation systems. More detailed discussion and examples of the use of the many available variations on these basic themes are given in Chapters 3 to 7. Chapters 8-11 present a number of complete (automatic and semi-automatic) line drawing interpretation systems.

To perform research in drawing interpretation is to seek and evaluate new processes, representations and system architectures. The set of methods and approaches discussed here should not therefore be thought of as fixed forever, but as a view of the area at the time of writing. This being said, it should be pointed out that the methods discussed here have been in use in one form or another throughout the period of development of line drawing interpretation technology. While they may, and hopefully will, develop as time goes on, it seems unlikely that these techniques will be entirely replaced. Those interested in commissioning a line drawing interpretation system would therefore be well advised to become conversant with both present approaches to drawing interpretation and the concepts which underpin them.

2.2 Five Stages of Line Drawing Interpretation

The precise form of the drawing interpretation problem varies considerably from application to application. Any one of a number of interpretation strategies may provide an acceptable solution to a given problem and each strategy may be implemented using diverse image analysis techniques combined within one of a range of system architectures. For all this variety, line drawing interpretation can still usefully be thought of, in broad terms, as the successive transformation of graphic information from one representation to another at a higher level of abstraction. At the lowest level is the original paper drawing: all the information required to perform the task at hand is contained, much of it only implicitly, within this document. Subsequent processes generate representations making explicit increasingly more abstract properties of the drawing and the information contained therein. Five such transformations are commonly recognised, these are:

1. scanning of the paper drawing to obtain a raster (binary or grey level) image;

2. vectorisation of the image to produce a representation of the input drawing in terms of simple graphical primitives;

3. extraction of universal drawing constructs (e.g. CAD entities or cartographic objects) from the vector and/or image representations;

4. extraction of specific 2D drawing objects described in, for example, a CAD or GIS library. This process typically also determines the parameters of each object and creates "scenes" by making explicit the geometrical and other relations between extracted objects;

5. reconstruction of a three-dimensional description of the item(s) depicted in the original drawing.

Assuming a simple progression from one stage to the next, line drawing interpretation may be summarised by figure 2.1.
  Scanning is the first step in any drawing interpretation methodology. A wide range of image acquisition and document scanning equipment, producing raster images in a wider variety of file formats, is now available commercially. Scanning technology, systems and the more commonly used image file structures are reviewed in Chapter 3. From a drawing interpretation perspective, one of the more important questions to ask of a scanner is whether the images it produces are grey level or binarised. Most grey-scale scanners distinguish at least 256 image intensity values: pixels corresponding to black areas of the scanned drawing will contain zeroes, those arising from white regions will contain 255. This convention may be reversed; black pixels may contain 255 while white pixels are denoted by zeroes. In either case

intermediate pixel values are a (roughly) linear function of shade of grey; mid-grey regions should generate pixel values of around 128, for example. A binary image can distinguish only two grey levels; 0 denotes black and 1 white, or vice versa. Binary images occupy much less space than grey level images, but contain less information; the ability to draw fine distinctions between shades of grey is clearly lost.

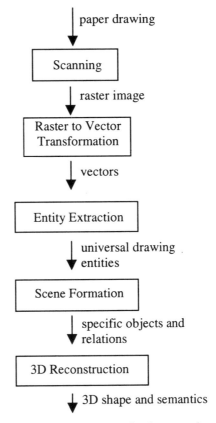

Figure 2.1. The five stages of line drawing interpretation.

The process by which a grey level image is binarised is commonly known as thresholding. Each pixel in the grey level image is compared to a threshold value; if the pixel intensity is greater than the threshold the corresponding pixel in a binary image is set to 1. Should the grey level be less than or equal to the threshold the binary image receives a zero. Thresholding is ubiquitous in image processing and analysis, in the current context the goal of thresholding is to convert an input grey level image to a binary array in which black ink-marks are distinguished from plain white paper. This process is discussed in detail in Chapter 4.

Thresholding is incorporated, in some form and at some stage, into all line drawing interpretation systems. While the ideal drawing comprises uniform black ink on uniform white paper, real drawings are never perfectly binary. Any process which inputs a grey level drawing and outputs a description of that drawing's content must therefore distinguish ink from clean paper at some point. Thresholding does not appear as a separate operation in figure 2.1, however, because it may be performed at almost any point in the transformation. Perhaps the most common approach is to incorporate thresholding into either the scanner hardware or raster to vector transformation process. There exist systems, however, which integrate thresholding into entity extraction and scene formation (Chapter 4). For the present it is enough to note that thresholding is an important, though implicit, component of the transformation shown in figure 2.1

Given a raster image, the raster to vector transformation (often referred to simply as vectorisation) generates a description of the drawing in terms of an unstructured set of straight line segments and other simple graphical primitives. While the results of these operations can be displayed on a CAD or GIS terminal, perhaps as a backdrop to manual input, their use is otherwise limited. Vectorisation methods and the processes needed to support them are considered in Chapters 5-7.

Entity extraction clearly has different connotations for different drawing disciplines. The symbols used in electrical schematics, for example, are usually small, well-defined patterns similar to special characters. In contrast, the symbols found in mechanical drawings are complex and variable groupings of graphical primitives that do not fall so easily to the pattern recognition techniques that have been successful elsewhere. CAD descriptions are normally expressed in terms of larger constructs and most CAD/CAM applications assume this higher level of representation. Universal graphical primitives such as arcs, circles, ellipses, text, arrowheads, dashed and chained lines, etc., must therefore be identified. Text recognition may also be incorporated at this point; if individual characters and/or words are not recognised, text should at the very least be separated from graphics during this process. The major cartographic objects (line types, symbols and others) should be recognised at this stage. A more detailed examination of entity extraction from images of engineering drawings and maps is given in Chapters 8 -11.

Scene formation involves recognising specific objects or object classes in a set of universal drawing entities. A scene formation system might, for example, identify standard components like bolts and brackets. It might also determine the sizes of these components and make note that a given bolt is being used to attach a specific bracket. Scene formation must also perform the rectification needed to make the graphical primitives consistent with the textual and symbolic information included in the drawing. For example, as mechanical drawings are not necessarily drawn to scale, size information must be read from dimensioning and propagated throughout the drawing. Cartographic scenes are based on relationships between objects. Those relationships must be extracted and the required objects should be grouped to form a scene. This is generally a very complicated task, which may involve complex spatial reasoning and falls beyond the scope of the present book.

The final step in this general drawing interpretation process involves the combination of views, sections and details supplied textually and via draughting convention to construct a 3D model of the drawn object(s). For complex drawings this may require knowledge of the type of object under consideration. Several research workers have addressed this problem though the resulting systems are usually limited to idealised drawings of simple artefacts. This is partly due to a lack of complete solutions to the intermediate entity extraction and 2D geometry reconstruction problems; little real data exists that can be used as input to a developing 3D model reconstruction system.

Three-dimensional reconstruction is of considerable interest and may be valuable in some applications. It should be noted, however, that in many situations entity extraction, perhaps incorporating a small element of scene formation, might be all that is required. Line drawings are more a form of communication than a direct reflection of physical reality. It is therefore not clear that full recovery of the 3D shape of drawn objects is always necessary. When examining a technical drawing, do mechanical engineers form three-dimensional mental models of the object it depicts? Or do they read off the information they need from the two-dimensional pattern of lines they see using their knowledge of drawing conventions? It seems likely that the information required to perform many tasks can be obtained without explicit recovery of the third dimension.

In practice, each of the steps listed in figure 2.1 would probably be followed by a significant amount of interactive post-processing. Even the early stages of the interpretation process can be expected to generate errors, which can, at present at least, only be corrected by a human operator. After thresholding, raster editors provide tools to add and delete black pixels and allow the user to clean up the binary image prior to vectorisation. Vector editors allow similar modifications to be made to the vectorised drawing, supplying tools to add/delete line segments and join or separate vector chains. Towards the end of the processing chain any errors made become more complicated and require more work from the operator using more powerful interactive tools. The design of these tools is a significant task in its own right, and may be as complex as the automatic component(s) of the system.

2.3 Intermediate and Target Representations

Figure 2.1 depicts, in idealised form, the major components of a line drawing interpretation system. So far we have discussed only the processes comprising drawing interpretation. Attention now turns to the six levels of representation involved: the initial line drawing, raster images, vector representations, universal drawing entities, 2D objects and 3D descriptions. The types of initial paper drawing that might be encountered, and their characteristics, were described in Chapter 1.

2.3.1 Raster Representations

Raster image representations may be used to describe grey level or binary images (figure 2.2). Most document scanners incorporate thresholding techniques that are quite adequate for many drawings. The digital image presented to the interpretation process is therefore often binary, though the scanned image is usually available in grey level form. This is useful when the paper drawing is faint or lacks definition; the thresholding methods built into modern scanners quite reasonably assume the scanned document to be of fair to good quality.

Paper line drawings are often very large, generating potentially unmanageable amounts of raster data. The technical difficulties caused by huge raster files may be alleviated in a variety of ways. One approach is to divide large raster files into parts or "frames" which after some degree of interpretation are joined back together. An example of such processing is shown in figure 2.3; a special co-ordinate system for frames is used. Each frame is processed individually and, for example, the end points of vectors extracted. The resulting data is then expressed within a universal co-ordinate system. End points in each frame are matched with the end points found in adjacent frames and combined to identify lines crossing frame boundaries. There are some difficulties in processing frame borders in this way (matching can be a non-trivial operation) and the additional image cutting and frame joining operations can consume significant computational resources.

Figure 2.2. A binarised engineering drawing image.

Another approach is to store and process in a special buffer a limited number of adjacent image lines: a "stripe". As interpretation progresses this stripe is moved up and down the image as necessary (figure 2.4). Striping is similar in some respects to the image partitioning method outlined above. The important difference is that in the former technique the image is divided into a fixed set of regions while in the latter the position and width of the stripe may be varied under the control of the interpretation system. The major advantage of this approach is that only the comparatively small image stripe is stored in the computer's memory at any given time; machines with quite severely restricted memory resources may therefore process images of any size. A primary drawback of striping is the additional burden imposed upon the interpretation system by the need to determine and acquire the image stripe needed to support a given operation. If the operation to be performed involves accessing many widely separated areas of the image, tracking a line which spirals into the centre of the image would be an extreme example, the overheads incurred in constantly modifying the stripe may render the approach impractical.

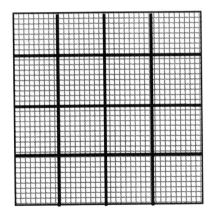

Figure 2.3. Static partitioning of large raster images.

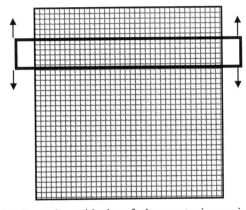

Figure 2.4. Dynamic partitioning of a large raster image via striping.

A third, and now more commonly used, solution to the problem of large raster files is to employ an image representation which records the raster data more efficiently than does a simple two-dimensional array. A wide variety of image compression techniques exist [38], all of which exploit some form of redundancy in the classic array representation to produce a coded image occupying significantly less memory than the original. It is clearly vital that the drawing interpretation techniques to be used are able to operate upon the compressed image, without decoding it, otherwise all the advantages of compression will be lost. Though this may seem obvious it may not be so easy to achieve; many image processing and analysis techniques rely heavily on the traditional array representation and either cannot function, or cannot function efficiently, given compressed image data.

A detailed and comprehensive survey of image compression and coding techniques is beyond the scope of the current text. Some of the more commonly used methods are, however, outlined in Chapter 3.

Regardless of the coding scheme employed, each raster image must be assigned a co-ordinate frame within which the position of a given pixel may be expressed. Two schemes are in common use (figure 2.5). The first places the origin at the top left of the image and gives pixel position in (row, column) format. This is perhaps the most frequently used approach. The second sites the origin at the bottom left of the raster array and uses (x, y) notation, where x is measured positively to the right and y positively in the vertical direction. Some computer vision systems use (x, y) notation with the origin at the centre of the image, though this method is hardly ever employed in line drawing interpretation work.

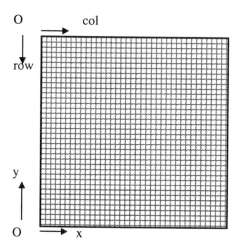

Figure 2.5. The two most commonly used image co-ordinate systems.

2.3.2 Vector Representations

While raster-based representations (Section 1.3) of line drawing images can be produced in a number of ways (see Chapter 3), vector representations are usually the result of an explicit vectorisation process. The name implies a description of connected components only in terms of straight line segments, though in practice curvilinear segments may be used. This being said, straight lines form the basis of most vector representations and we shall continue, as do the majority of those writing in this area, to use the term vector to refer to curved as well as straight segments.

Vector representations of line drawings usually take the form of an ASCII file containing a number of descriptive statements, each written in a simple, standard form and describing one specific line or curve segment. One might find, for example, that the end points of each straight line are recorded. Alternatively, one end and a vector giving the relative position of the other might be stored. A circular arc might be described by noting its centre, radius and end points in, say, clockwise order. Details will vary from author to author and system to system. It is common, however, to assign each vector a label specifying its type (straight, circular, etc.) and an identifying number or code. A vector file entry describing a straight line from pixel (15, 15) to pixel (100, 100) might, therefore, be written:

S 1 15 15 100 100

Here S would indicate a straight line, 1 is an identifier allowing subsequent processes to distinguish one entry from another and the remaining values give the co-ordinates of the end points. Similarly,

C 24 50 100 100 100 50 50 100

might mean that circular arc number 24 is of radius 50 pixels, has its centre at pixel (100, 100) in (row, column) notation, starts at pixel (100, 50) and extends clockwise to pixel (50, 100).

There is clearly some redundancy in the above representation; the radius may be determined from the centre and an end point. While this results in a slightly inefficient description, it may make the representation easier to use and so increase the speed of subsequent operations. In general, the advantages and disadvantages of a particular coding scheme must be evaluated in the context of the system within which it is to be employed. A now classic, and quite detailed, review of the vector representations in general use in image processing and analysis is given by Freeman [39]. Those familiar with standard CAD formats such as DXF will recognise components of those formats here.

A typical vector file will contain several hundred segment descriptions. These are usually written into the file in the order in which they are extracted from the image array. As a result, vector files are generally unstructured collections of distinct vector descriptions. While important relationships between vectors (connectivity, colinearity, parallelism, etc.) may be obvious to a human viewing the line drawing, these are not normally made explicit in vector files. The comparison and grouping together of vectors needed to identify these relationships is usually left to the entity extraction process.

One question remains: exactly what parts of the line drawing image do vectors describe? Each line on a drawing will appear in the raster image as an elongated region of dark pixels. Although the length of these regions is usually much greater, they are regions and do have a finite width; ten or so pixels is not unusual. Vector representations, however, often describe mathematically ideal lines, which have zero width. The question therefore becomes; where within the image region corresponding to a drawn line should the idealised vector lie? Two answers are commonly given (figure 2.6).

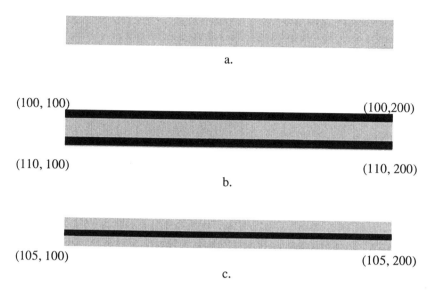

Figure 2.6. a) The image region corresponding to part of a drawn line, b) A contour representation; vectors lie on the region boundary, c) A skeleton representation; vectors lie on the median line. Vector end points are given in (row, column) format.

The first approach is to produce a contour representation. The boundaries of the inked region are located by thresholding, then described using vectors. As a result, each image line has two associated vectors (figure 2.6.b). For drawings comprising only simple lines, like that shown in figure 2.6, this form of representation may be considered too costly in terms of storage space. Many of the lines to be found in engineering and technical drawings, however, are annotated in some way: arrows are perhaps the most common example. Contour-based vector representations can be very useful when attempting to automatically identify annotated lines as the pattern of vectors obtained will reflect the annotation (figure 2.7).

Figure 2.7. Vectors obtained from the contour of an arrow reflect the shape of the arrowhead, making its detection easier.

The second approach is to fit vectors through the centre of the inked region (figure 2.6.c). This may be done in a variety of ways, though perhaps the most common method is to apply a technique known as thinning to the thresholded image to first identify a one-pixel wide string of pixels lying along the region's centre or

Components of a Drawing Interpretation System

medial line. Line and/or curve segments are then fitted to this thinned data to give a vector representation. This type of description is often referred to as a line drawing "skeleton".

Skeleton-style vector representations often contain two types of data: vector segments and knots describing connections between vectors. This form of vector-based description is generally more structured than the previous one. It typically contains detailed information not only about segment locations but also about their characteristics and relations. Additionally, this form may include information regarding the segment's geometric class (straight line, circular arc, etc.) or curvature properties. As each pen stroke generates a single chain of vectors, not two as in the contour method, skeleton representations are reasonably compact (figure 2.8).

Figure 2.8. Skeleton-style vector representation.

One drawback, however, is that annotation and thick areas are often lost in the skeletonisation process. As a result, some systems attempt to achieve the best of both worlds by using a skeleton representation unless there is evidence for or an expectation of thick areas, in which case a contour-based approach is taken (figure 2.9). The construction of mixed contour/skeleton representations requires access to local estimates of line width. Techniques capable of providing this data are discussed in Chapters 5 and 6.

Figure 2.9. Mixed skeleton and contour object representation.

2.3.3 Universal Drawing Entities

If a line drawing is to provide an effective means of communication, some form of drawing standard or convention must be agreed. Drawing standards take many forms. All, however, share a common feature: each specifies a set of universal drawing entities from which acceptable drawings must be composed. Specific instances of a given entity may vary considerably. Crosshatched regions in mechanical drawings, for example, may be of any size and shape. They should all, however, comprise evenly spaced parallel lines which are slightly narrower than the lines denoting the physical outline of the drawn object and drawn at an angle of 45 degrees to the edge of the paper. Similarly, though there are various forms of dimensioning, all should include a leader line appropriately marked with arrowheads, two witness lines linking the leader to the outline of the object, and a piece of text adjacent to the leader which gives the dimensions. In ordnance survey maps, houses may be any shape or size but must appear as closed sequences of pen strokes. Similar conventions exist for other drawing types. Image representations couched in terms of universal CAD or cartographic entities provide high-level descriptions of the input engineering drawing or map which are suitable for direct entry into CAD or GIS systems. These representations are typically highly structured collections of graphical primitives.

Graphical primitives may usefully be divided into two classes: simple and complex. Straight lines, arcs, curves, circles are the most commonly used simple graphical primitives, though characters and special symbols may also be employed (figure 2.10).

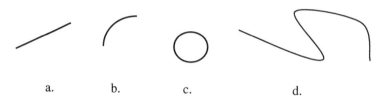

Figure 2.10. Simple graphical primitives: a) straight line, b) arc, c) circle, d) curve.

Complex primitives are obtained by grouping together simple ones. The members of these groups may be divided by areas of blank paper or have different parameters. Multiple simple primitives are united to form one complex primitive on the basis of some shared geometric or logical characteristics. Complex primitives describing dashed or chained lines may be formed from sets of collinear straight line segments, for example, while blocks of text may be obtained by grouping together characters and other compact, solid regions (figure 2.11).

For many drawing interpretation systems, the formation of complex from simple graphical primitives is only the first time that grouping processes are applied. Complex primitives representing the chained lines, text blocks and arrows in a mechanical drawing may later be combined to form still higher-level entities describing, for example, a specified dimension. Representations of individual houses extracted from an ordnance survey map may be grouped to describe the row of houses making up a terrace. In general, universal entity representations are hierarchically structured: simple graphical primitives are the base elements from which increasingly complex entity descriptions are constructed.

Components of a Drawing Interpretation System

Figure 2.11. Complex graphical primitives; a) arrow, b) dashed line, c) symbols, d) variable-width line.

Their natural hierarchical structure gives universal entity representations a flexibility that is very well suited to use in line drawing archives, CAD and GIS systems. The simple primitives stored at the lower levels of the hierarchy provide a detailed description of the drawing and allow simple graphics or printing systems to produce images of the drawing which are very similar to the scanned original. By focusing on the upper levels of the hierarchy and perhaps discarding the lower levels altogether, a more compact but less detailed representation may be obtained. This will be oriented towards more complex drawing interpretation tasks but only able to support production of a relatively poor image of the original drawing.

The precise nature of a universal entity representation may vary with the type of drawings to be input to the interpretation system and the tasks to be supported by its output. There will, however, be some similarities both between and within drawing disciplines. A typical entity-level description of a mechanical drawing might include centre and symmetry axes (represented by dash-dotted lines), hidden contour lines (dashed lines), solid areas (crosshatching), dimensions (thin lines with arrows, witness lines and so on) and textual annotations (figure 2.12). Similar entities appear in other engineering drawings. It is worthwhile stressing at this point that these objects will often be far from perfectly drawn; as shown in figure 2.12, for example, it is commonplace to find dimensioning which merges with the dimensioned object and unevenly spaced crosshatch lines which extend beyond the crosshatched region.

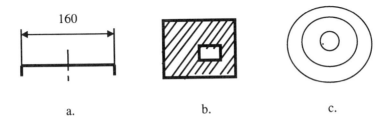

Figure 2.12. Engineering drawing entities as they might appear in a drawing; a) axis of symmetry, b) crosshatched area, c) a set of concentric circles.

Universal entity descriptions of maps are based on fairly standard cartographic objects. Examples of some such entities, which vary from country to country, are shown in figure 2.13.

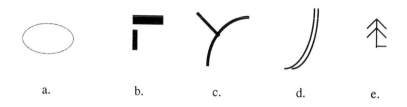

a.　　　　　b.　　　　c.　　　　d.　　　　e.

Figure 2.13. Examples of cartographic objects: a) isoline, b) buildings, c) rivers, d) road, e) conventional sign.

Each entity description must store object type and metrics and may record some semantic information. An object's type is determined by its graphical structure and implies a concrete object or object class on the original drawing. To determine type, the system must exploit knowledge of the global structure of the object and its components. This information can be represented in a variety of ways, from ad hoc rules of thumb to expressions in a formally defined language. Object metrics are determined from the object's disposition on the image and vary with object type. Line objects may be represented by their skeleton and width, symbols by one or two points and regions by their bounding contours. It is important to note that object metrics refer to the size, shape and other properties of the graphical primitives which comprise the entity. In contrast, the semantics of an entity reflect concrete characteristics of the real object. For example, an isoline's metric information specifies its image co-ordinates while its semantics define the height of the strip of land to which it refers.

Regardless of drawing type, the data structures containing universal entity descriptions will typically be connected together in some way to provide a coherent representation of the input drawing. This is done either physically by storing shared co-ordinates within different entity descriptions or logically by one entity making explicit reference to others. However it is achieved, a universal entity level description of an input line drawing is a rich data structure containing much useful information which may be employed in the execution of a variety of drawing-related tasks.

2.3.4 Two-Dimensional Objects

The construction of universal entity representations generally involves a steady progression from low-level, simple graphical primitives to higher level, increasingly more complex data structures. In the later stages of this process, when suitably detailed and complex representations exist, it may be both possible and desirable to cease building universal entities and begin to seek specific objects that may be expected to appear in drawings from particular application areas. In some

Components of a Drawing Interpretation System

mechanical engineering applications, for example, particular gearboxes and drive shafts might be commonplace. In other mechanical domains, perhaps within a company which designs and produces, say, compressed air storage and delivery devices, motor components may rarely or never be drawn. Draughtsmen operating in both environments will, however, produce drawings of their particular objects of interest using the same set of universal entities.

In map interpretation, 2D objects correspond to geographical scenes. Scene descriptions usually comprise several objects joined together by meaningful relations. Examples of cartographic scenes are shown in figure 2.14. Specific objects are typically represented in much the same way as high-level universal entities; they are two-dimensional descriptions of the drawing expressed in terms of simple and complex graphical primitives. The key difference is that a process of recognition has matched these descriptions to pre-stored descriptions of the specific gearboxes, shafts, conurbations etc. that are likely to be found in the drawing and added a label to the entity descriptions involved saying which object they depict. It is possible that representation of some two-dimensional CAD or GIS objects will require new and perhaps more complex relations between entities to be extracted and that this will generate a need for more complex kinds of control information, but the principles of entity description will usually remain valid.

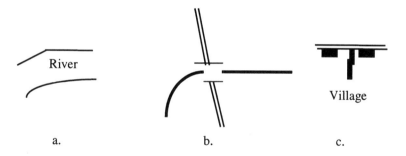

Figure 2.14. Example of cartographic scenes; a) river with signature, b) bridge, c) village.

2.3.5 Three-Dimensional Shape and Semantics

In some situations line drawing interpretation must provide a three-dimensional description of the drawn objects. It may be necessary, for example, to produce a relief map of an area depicted by ordnance survey maps or to extract from mechanical drawings the 3D shape of the surface of an object that is to be machined. In broad terms, three-dimensional representations may be obtained in either of two ways. In the first, semantic information contained in the entity-level description forms the basis of the reconstruction process. Here the graphical representation (structure, colour, etc.) of the object and its components provides the necessary information, perhaps with the help of special signs and symbols (signatures, texture elements etc.). Contour maps provide enough semantic information, in principle, to allow 3D relief maps to be obtained in this way.

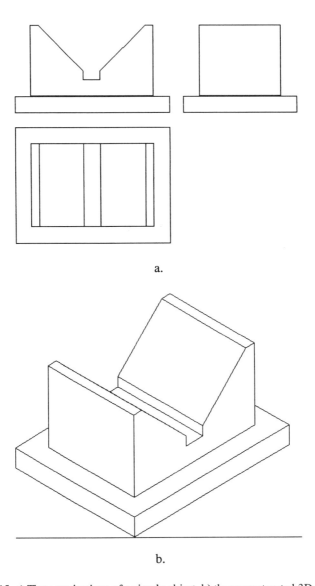

Figure 2.15. a) Three projections of a simple object, b) the reconstructed 3D model.

The second approach has more in common with techniques used in three-dimensional computer vision and is of greater relevance to engineering than cartographic domains. Here, two or more projections of the object of interest are considered. Each projection is a separate drawing showing the object from a different viewpoint. The six standard orthogonal views (front, top, right side, left side, back and bottom) or some subset of these are most commonly used [40]. A three-dimensional representation is obtained by combining these multiple views, i.e. noting which line in one projection denotes the same object feature as a given line in another projection (figure 2.15). If this correspondence problem can be solved the relative positions of the relevant object features can usually be recovered, though supporting semantic information may also be required, either to aid correspondence or to guide interpretation of the matched features.

Although specific two- and three-dimensional object descriptions are perhaps the most desirable drawing representations, these cannot yet be produced automatically in any but the most restricted application domains. One reason for this is that only now are the problems of universal entity extraction becoming well understood and reliable entity extraction methods being developed. It has been appreciated for some time that if systems producing these more advanced representations are to become a reality further theoretical development is needed, but without entity level data with which to work this development is hard to achieve.

A second reason for the current relatively poor performance of 3D reconstruction methods is that many of those working in drawing interpretation believe that explicit 3D descriptions may not actually be necessary. It could be that all the information needed for the engineer to solve his/her task is available in more accessible 2D representations. It may therefore be more productive in the long run to concentrate on developing principled and reliable methods of obtaining universal entity and 2D object representations, and to address the 3D reconstruction problem only when absolutely necessary.

2.4 System Architectures and the Role of A Priori Knowledge

By now it should be apparent that the interpretation of images of line drawings is a complex task: up to five major types of drawing representation may be involved, these may be connected by processes which range from simple thresholding to 3D reconstruction from multiple projections. Some of the activities appearing in figure 2.1 may be combined within a single process; it may be advantageous to combine vectorisation and entity extraction, for example, and design a single process which converts a raster representation into an entity-level description. Moreover, to function correctly, any process may be required to access and make efficient use of the available a priori knowledge. Prior knowledge might include anything from expected distributions of grey levels within the input image to constraints on the relationships between high-level two- or three-dimensional objects. It is not enough to develop efficient drawing representations and effective, but independent, thresholding, vectorisation and entity extraction processes; careful thought must be given to the ways in which these individual parts might be combined to produce a coherent whole. The design of a drawing interpretation system's architecture is at least as important as the design of its components.

Current approaches to line drawing interpretation can generally be classified as either bottom-up or top-down [41]. The bottom-up approach is characterised by an emphasis on the analysis of small groups of connected or otherwise physically closely related pixels and relies upon data-driven, local processing. Bottom-up interpretation systems tend to start with the image and move towards abstract, entity-level descriptions. In contrast, top-down approaches concentrate on the relationships among graphical primitives, objects and scenes. Systems built around this type of architecture typically begin with some description of the entities they expect to find and proceed by seeking evidence for the presence of those entities in the input drawing. Such systems therefore rely more on model-driven, global processing. At the time of writing bottom-up approaches remain the most commonly used, particularly for lower level interpretation tasks such as map vectorisation. Top-down approaches are usually referred to as knowledge-based because they use a priori knowledge to guide object recognition.

2.4.1 Sequential Architectures

The simplest, and most commonly used, bottom-up line drawing interpretation system architecture is the classic sequential structure. Systems based upon this model comprise a linearly ordered set of independent processes: each receives the output of its predecessor and passes its own output on to the next process in the chain. The drawing image is input to the first and the interpretation emerges from the final process. Figure 2.16 shows the structure of a drawing interpretation system [13] based on this design.

Implicit in the sequential architecture is the assumption that drawing interpretation can be modelled as a set of independent processes whose interfaces can be fixed at compile-time. That is:

- the goals of each process can be precisely defined without reference to any other process in the system;

- all the possible interactions between the system's components are precisely known when the system is created.

When this is true there are several benefits. Each process can be designed and implemented independently of all the others, perhaps by a different member of the development team. The simplicity of the architecture means that there is unlikely to be any confusion as to the requirements of each component. As long as the requirements of a particular processing stage are met, any image processing/analysis technology can be used in its implementation. If, for example, a new vectorisation algorithm becomes available it should be a simple matter to unplug the previous vectorisation process and insert the improved version. Each component performs a well-defined operation and provides a limited number of comparatively simple interfaces to other modules. Components may, therefore, be used outside the system for which they were designed, perhaps independently or in systems performing other drawing interpretation tasks.

If the assumption underlying the sequential design holds, the architecture is attractive. What is not clear, however, is how often this assumption actually is true. In many situations a strictly bottom-up style of processing comprising a rigidly ordered set of filters is obviously neither powerful nor flexible enough for the task at hand. The approach has several problems.

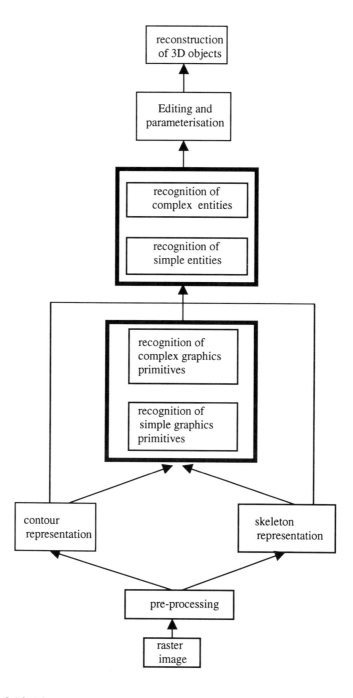

Figure 2.16. A bottom-up, sequential engineering drawing interpretation system.

First, the filters employed are often too general; few are tuned in any way to the special problems of drawing conversion. For instance, a common approach to vectorisation is through thinning followed by tracking of the skeletonised image. When this technique is applied variations in line width often generate short spurs that must be eliminated. Careless removal of short line data, however, can result in the loss of vital information. It is important, for example, not to discard spurs generated by thinning the arrowheads on a dimension line. This sort of mistake, which could have severe consequences for the interpretation of the image, can only be avoided by incorporating knowledge of the structure of the drawing. It is not practical, however, to build into each separate filter all the knowledge that it may require. Rather the individual processes should be integrated within a suitable knowledge-based environment.

Consider thresholding, for example. Situations often arise in which one can only decide that a faint line is important after performing a detailed and high-level interpretation of the surrounding drawing/image. In such cases thresholding should ideally be reapplied under the control of higher processes which can tune its parameters to perform a realistic, directed search for the faint structure. Without the ability to adopt this type of flexible processing strategy important decisions are made too early. In systems built around the sequential architecture most of the processing is done without reference to the source image or other lower-level representations. Performance is therefore crucially dependent on all the earlier transformations being accurate. Given the poor quality of many line drawings this extremely high degree of reliability is unlikely to be attainable. Although sequential architectures have their advantages and can produce workable systems, it seems unlikely that this style of processing will provide a generic solution to the drawing interpretation problem. Other architectures should at least be considered.

2.4.2 Blackboard Architectures: The Expert System Approach

Reading an engineering drawing or map is a skilled task. Experienced mechanical engineers can extract much more information, much faster, from piece-part drawings than is possible for an untrained member of the general public. Maps also provide a richer source of information to the trained eye. One might go so far as to argue that interpreting certain types of line drawing is an expert task, in much the same way that interpreting a patient's symptoms and test results to make a medical diagnosis is considered to require learned expertise.

The construction of so-called "Expert Systems" in various domains, including some capable of (limited) medical diagnosis, has received much attention over the last thirty years. The study of expert systems was originally considered part of the more general field of Artificial Intelligence, though in recent years the two topics have grown apart. There is now a wide variety of books, journals, conferences and societies dedicated to the continuing development of expert or knowledge-based system technology. For the present we shall simply contrast the expert system and sequential approaches by introducing the classic expert system architecture: the blackboard.

The aim of the blackboard architecture [42] is to provide a flexible framework for the construction of knowledge-based systems. The approach is inspired by the idea of a group of collaborating human experts, all standing around a real blackboard. As

Components of a Drawing Interpretation System

an individual expert generates ideas he/she makes them public by writing on the board. Those ideas are then accessible to all the other experts, who can present arguments for and against or propose modifications to them, again by writing on the blackboard. This process continues until some form of agreement is reached, at which point the solution to the problem at hand is available from the board.

In a blackboard architecture expert system (figure 2.17) the human experts become "knowledge sources", programs which embody the system's knowledge of some specific area or sub-task. The blackboard becomes a set of globally available data structures. Each knowledge source can access the blackboard more or less at will. Some conflict resolution system is required, however, to ensure that two knowledge sources do not try to write to the same area of the blackboard at the same time or to add contradictory statements to the developing solution.

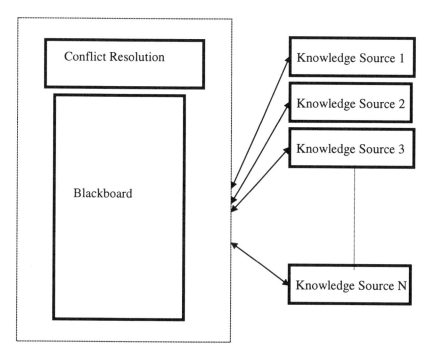

Figure 2.17. The blackboard architecture.

The strength of the blackboard architecture lies in its flexibility; each knowledge source adds what it can when it can to the developing interpretation. As a result, a given system may operate in a bottom-up fashion in one situation and top-down in another. Indeed, it is often impossible to predict how a blackboard system will behave. This rather extreme level of flexibility obviously brings its own problems; an unpredictable system is very hard to maintain, develop and market.

Despite these drawbacks, a number of successful image interpretation systems have been built around the blackboard model. These may be broadly divided into two classes. The first exploits the flexibility of the blackboard architecture by using it as an experimental environment within which an interpretation strategy may be developed. The idea here is to construct prototype systems using the blackboard approach and then, once the system design has stabilised, to convert the blackboard prototype into a production system based upon a more restrictive architecture. The second approach is to maintain the blackboard architecture throughout the system's entire life-cycle, but to place restrictions on the interactions between knowledge sources. This leads to a system which is more powerful than a sequential design but more manageable than a full-blown blackboard. We shall return to this topic in Chapters 8-11.

Chapter 3
Document Image Acquisition

3.1 Scanning Devices

In the wider field of image processing and analysis the most common source of grey-level images is the CCD (Charge-Coupled Device) camera. This quite closely resembles a standard photographic camera; a lens at the front of the unit focuses light from the viewed environment onto a light sensitive surface within the camera body. In photography that surface is a section of photographic film. In CCD cameras, film is replaced by a rectangular array of light sensitive cells. When the shutter opens, each cell generates an electrical signal proportional to the amount of light falling upon it. These signals are fed out of the camera and into the computer via special purpose hardware which converts the analogue signals generated by the CCD array into a digital raster image.

CCD cameras could in principle be, and very occasionally are, used to acquire images of line drawings. The large size of most drawings, however, causes problems; the first is poor resolution. CCD cameras typically employ arrays of around 512 x 512 charge-coupled cells, generating a 512 x 512 pixel image. Television resolution is 768 x 576 pixels. Larger CCD arrays are available, but these quickly become expensive. If an entire A0 drawing is imaged by a 512 x 512 cell array, each pixel will represent an area of the document approximately 3mm square. This, perhaps, does not sound too bad. Most engineering drawings, however, comprise significant numbers of lines with widths of around 0.5mm. At a resolution of 3 mm per pixel, lines this thin may simply not be visible in the raster image. Drawings can also contain very closely spaced lines. At low resolutions these will be merged together and important detail lost.

A second problem is foreshortening. If the CCD array is not parallel to the drawing when an image is acquired, that image will be distorted by perspective (figure 3.1). While there exist camera calibration techniques (e.g. [43]) which can determine the relative orientation of the drawing and camera, these are computationally quite expensive. The use of such techniques in drawing image acquisition, moreover, seems excessive. Camera calibration is an accepted component of computer vision systems; most three-dimensional objects may legitimately be imaged from a large number of viewpoints and it is useful, if not always necessary, to recover as much information about viewing conditions as possible before beginning to interpret the image. If the object to be imaged is a two-dimensional planar sheet of paper only one viewpoint, shown in figure 3.1.a, is really desirable. The flexibility provided by camera calibration is not needed. It makes more sense to construct imaging equipment which can capture images from this, and only this, viewpoint.

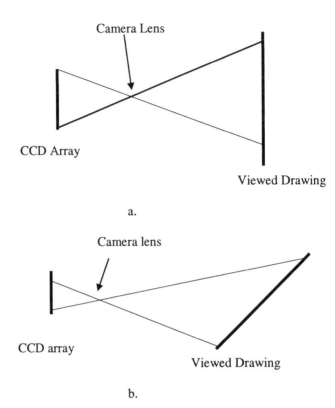

Figure 3.1. a) If the viewed document and CCD array are parallel, the resulting raster image is a direct representation of the drawing, b) when the drawing and CCD array are not parallel, perspective effects distort the image.

Document scanners are specifically designed to acquire images of planar objects. They are intended to hold the input document in a fixed spatial relationship to a CCD array, and so are able to acquire images from only a single viewpoint. Lighting conditions are also fixed; the page is held in a closed compartment and illuminated only by internal lamps. Imaging conditions are therefore carefully controlled. As a result, even the cheaper scanners tend to produce more consistent, better quality document images than do systems based upon calibrated CCD cameras.

Large format scanners are commonly used to input engineering drawings and maps. Some can handle colour documents, others produce only grey level images. Some can scan a coloured document and split the resulting image into a number of different files, each containing the components drawn in a particular colour. Many companies produce scanners and it is possible to find and buy equipment satisfying practically any reasonable requirement. As a diversity of scanner designs exists we cannot describe all of them here. One can, however, identify three basic scanner types: flatbed, overhead and sheetfed.

Flatbed scanners are the most mechanically sophisticated and so are usually more expensive. In flatbed designs the document is fixed on a flat transparent bed, information side down. Scanning is achieved by moving an optical reading head located under the document. This head comprises a linear array of CCD elements, so as the array is swept over (or more accurately under) the document an image is acquired (figure 3.2).

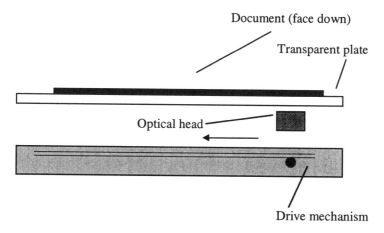

Figure 3.2. Schematic diagram of a flatbed scanner. Many variations on this theme are available commercially.

In overhead scanners the optical head is located above the document, which is again fixed in place. Scanning may be performed in a number of ways (figure 3.3). The reading element may be two-dimensional (figure 3.3.a), acquiring an image as would a suitably placed CCD camera. Alternatively, it may be linear. A linear CCD array may either be moved over the sheet (figure 3.3.b), as in flatbed machines, or remain fixed while a rotating mirror focuses successive linear segments of the document onto it (figure 3.3.c). The design decision linking flatbed and overhead scanners is that in both cases the document is fixed in place throughout image acquisition. The main advantage of fixed document scanners is the high resolution of the images they produce.

In sheetfed scanners the document moves while the (linear) reading head remains stationary (figure 3.4). These devices are typically cheaper than other designs but tend to produce lower resolution images. The need for a mechanism capable of moving the page also tends to restrict the size of document that can be passed through sheetfed scanners. As smooth, constant velocity movement of large documents is quite difficult (large sheets tend to wrinkle as they are fed in) most sheetfed scanners are limited to smaller (e.g. A4 or, at most, A3) pages. In the right circumstances, however, sheetfed scanners are ideal for high speed, low resolution input of mass data. When large drawings must be input, drum scanners (figure 3.5) may be used. Here the document is laid on a rotating drum which carries it past a linear sensing element; each rotation of the drum generates an image. As flatbed and overhead scanning technology becomes cheaper, drum scanners are becoming less common. They are still, however, quite widely used for the input of maps.

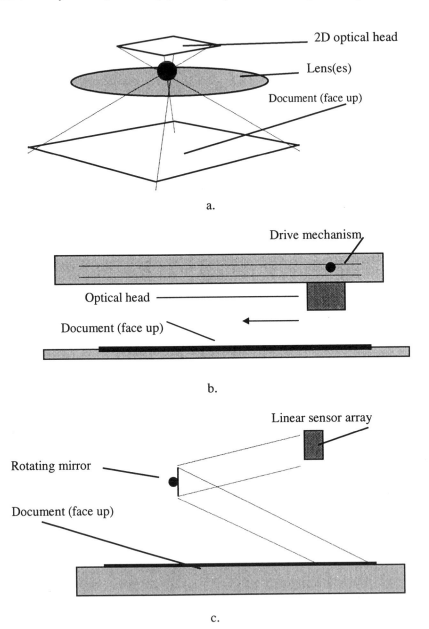

Figure 3.3. a) Schematic diagram of an overhead scanner with a 2D reading head, b) schematic diagram of an overhead scanner with a travelling linear head, c) Schematic diagram of an overhead scanner incorporating a rotating mirror.

Document Image Acquisition

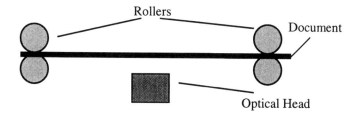

Figure 3.4. Schematic diagram of a sheetfed scanner.

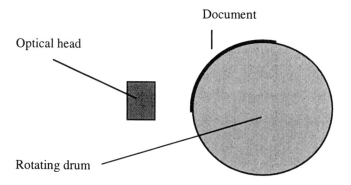

Figure 3.5. Schematic diagram of a drum scanner.

Regardless of basic design, a number of important characteristics should be taken into account when selecting a document scanner for a particular line drawing interpretation task. These are:

1. The range of drawing sizes that can be accommodated. Cheaper scanners input only A4 documents and are mainly used to support optical character recognition. Most line drawing interpretation systems, however, require high quality images of A0 sheets.

2. The range of media types that can be input. Many scanners can deal with any document medium including paper, photographic film, polyester film, linen and vellum.

3. Scanner resolution. This is variable on most modern scanners and is typically expressed in dots per inch (DPI) or pixels per mm. Scan resolution typically varies from 100 DPI to about 2000 DPI. For line drawing input, a resolution of somewhere between 300 and 600 DPI is usually the most suitable. If the chosen resolution is very low (e.g. 100 DPI) drawing objects will appear disconnected

in the image. If, on the other hand, scan/image resolution is very high (e.g. 1000 DPI) each penstroke will generate a thick band of maybe 10-15 dark pixels. This can greatly increase subsequent processing time. In general, resolution should be chosen after consideration of the contents of the initial document.

4. Scanning speed. This is usually measured in inches or centimetres per second and typically varies from 1 to 2 in/s. Many scanners provide a high speed scanning option, though this is usually achieved at the expense of image resolution.

5. Image enhancement capability. Many scanners now incorporate image enhancement processes that allow them to glean information from poor quality, weak blue-line and/or sepia drawings. Thresholding and simple noise reduction operations like the deletion of small black regions may also be performed automatically within the scanner.

6. Grey level option. Although automatic thresholding and enhancement can be very valuable, poor quality line drawings may require high quality or application-specific algorithms that will not be built into a low cost, general purpose scanner. The ability to output unprocessed grey level images, rather than just binarised data, can therefore be crucial.

7. Choice of output format(s). Upwards of 100 image file formats are currently in use. The best for a given application depends upon the context in which the work is to be done (see below).

8. Interfacing and control software. A variety of hardware interfaces may be used to link the scanner to the processor on which the interpretation system is implemented. Commonly used examples are Versatec Greensheet, DEC DR-11 and SCSI. As one would expect, an interface must be selected which provides easy connection to the remainder of the system at reasonable cost. Most scanners come complete with a software library intended to support communication between the scanner and host computer. A key question is whether this allows remote, and perhaps automatic, control of the image acquisition process. Some systems permit parameters such as scan area, resolution and contrast level to be controlled from the host computer. Others require these to be set by an operator via a keypad built into the scanner unit.

3.2 Image Coding

One way to overcome technical difficulties caused by the large size of many paper drawings is to employ a coding scheme that represents the image using less memory than would be required by the full pixel array. These coding methods are often referred to as image compression techniques and are said to produce compressed images. Primary applications of compression techniques are in image transmission and storage. Image transmission arises in, for example, remote sensing via satellite or aircraft, broadcast television and facsimile transfer. Images are compressed to reduce the body of information to be transmitted and so minimise transmission time. Image storage problems are ubiquitous and may involve educational and business documents, medical and satellite images, motion pictures, etc. A comprehensive review of image compression is beyond the scope of the current text; thorough discussions of the area are to be found in [44, 45]. We shall restrict ourselves to consideration of a few techniques commonly employed in line drawing interpretation. Line drawing interpretation systems do not usually rely significantly on bulk storage or transmission of grey level raster files. The initial image may, however, be sufficiently large as to necessitate some form of compression.

Scanned line drawings are generally held in binary form. Binary drawing images contain mostly white pixels interspersed with smallish black regions denoting characters, lines and other connected components. In this situation, several approaches to compression present themselves:

1. record only transitions between black and white pixels;

2. skip white pixels, recording only the black ones;

3. use pattern recognition to label black areas, then store only the labels and their locations.

As the aim of compression in line drawing interpretation is to support pattern recognition and similar processes, it is perhaps not surprising to find that most drawing compression techniques are based on approaches 1 and 2 above. The methods most frequently applied to line drawing images are run-length encoding, quadtrees, chain codes and various boundary representations.

Run-length representations record, in any of a number of ways, the locations of black/white boundaries and the colour of the pixels between them. Each line (or raster) of an uncompressed image comprises a number of adjacent strips of pixels, each strip containing only one grey value. These strips, or "runs", may be thought of as the building blocks of the image. Each raster of a binary image is completely specified by the lengths of successive runs and the colour (black or white) of the first. A common variation of the standard run-length representation is the so-called "modified run-length code" in which every line is represented by an ordered list of x co-ordinates of the black (or white) pixels which have white (or black) pixels on their left. This representation has been successfully used in one of our own vectorisation systems [45].

Quadtrees operate by dividing the binary image into square regions containing only black or white pixels. Assuming that the image is neither completely black nor completely white, it is initially divided into four quadrants. Then, any quadrant that contains both black and white pixels is further divided into four sub-quadrants. This recursive process continues until each sub-quadrant is either completely black or completely white, generating a tree structure; hence the name. Any quadtree may be represented by a unique character string comprising 3 basic symbols: b (black), w (white), and g (grey). Each g is necessarily followed by four symbols or groups of symbols representing the sub-quadrants of the grey (mixed black and white) area it represents. Quadtrees are more efficient from a data compression point of view, though the representation makes computation of object parameters such as perimeter, etc. more difficult [44].

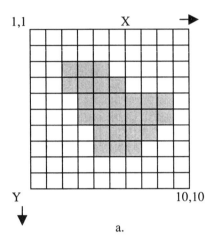

a.

Start Colour	Run lengths		
W	10		
W	10		
W	2	3	5
W	2	4	4
W	3	6	1
W	3	6	1
W	4	4	2
W	4	3	3
W	10		
W	10		

b.

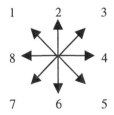

Direction values for chain code starting at pixel (3,3):
4455446778821212

c.

Figure 3.6. a) a simple binary image object represented b) in run-length form and c) by chain code.

Document Image Acquisition 53

Chain codes are used to represent black/white boundaries and/or thin (one pixel wide) lines. They record the direction of movements from one pixel to another during tracing of a boundary or line. Typically, the chain code comprises the image position of a start pixel followed by a string of direction codes. In the standard chain code eight distinct directions are identified (figure 3.6); this may, however, vary.

Boundary approximation techniques are closely related to chain codes. Here, object boundaries are approximated by polygons whose vertices make up the (compressed) representation of the object. Numerous boundary approximation techniques exist, some of which will be discussed in Chapter 6. Chain codes and boundary approximations are commonly used, but awkward when the objects to be represented contain interior holes. Additional information is then required to relate interior and exterior boundaries.

Run-length encoding is perhaps the most commonly used method of compressing line drawing images and a number of interpretation techniques relying on this representation [e.g. 46-48] have emerged as a result. Very few line drawing interpretation tools operate on quadtrees; although compact, the representation is difficult to work with. Chain code and boundary representations are frequently used in drawing interpretation, but at a later stage, after vectorisation. Figure 3.6 shows how a simple binary image object would be represented by some of these different techniques.

3.3 Image File Formats

Once scanning (and maybe compression) is complete, the image must be stored in a form in which it can be loaded into a developing drawing interpretation system, archived or conveyed from one site to another. Choice of image file format is usually determined by factors such as the intended use of the image, its size, whether or not it is to be compressed and what auxiliary information is to be stored along with the grey level data.

In 1993 Advanced Imaging magazine commissioned a review of file formats with the aim of identifying the more popular and widely used conventions [9]. The results of that review remain valid today and are summarised in Table 3.1. The TIFF (Tagged Image File Format) format was found to be the most popular. TIFF was developed in 1985 for use in desktop publishing. Its aim was to help prevent the introduction of competing proprietary standards by offering a fairly universal level of functionality, which has made it a relatively complex format. While it was and probably still is the most commonly used convention, TIFF has only partly met its target. Over 100 variations on the TIFF standard were in use in 1993, a situation which has changed little in the interim, causing some confusion as to what constitutes a "TIFF file".

The GIF format was developed by Compuserve to provide a hardware independent way of exchanging colour image files. It is now widely used in PC applications and to store images on the world-wide web. GIF probably commands a higher market share than it did in 1993, but it is unlikely to have overtaken TIFF in popularity.

VIFF was developed for the Khoros image processing package. As well as supporting Khoros, it was intended to ease the exchange of imagery between institutions. Khoros is widely used in academic image processing laboratories and in this sense VIFF has been successful. It seems likely, however, that usage of VIFF to store line drawing images has declined since the Advanced Imaging review was commissioned.

Some formats were specifically developed for in-house applications. Targa (or TGA), SunRaster and SGI were introduced by Truevision Inc., Sun and Silicon Graphics respectively for use on their workstation and video graphics products. EPS originates from Adobe and is different in that it is really a page description language. EPS only supports images because they are commonly found embedded in larger publications. FITS was developed to service the specialist needs of the astronomical community. The ZSoft Corporation's PCX and its variant PCC (PC Clip art) are image file formats developed specially for IBM PC applications [49]. PBM was originally developed to allow bitmaps to be sent by electronic mail systems unable to handle pure binary images.

Format	Usage (%)
TIFF	48
GIF	33
VIFF	30
PBM	27
Home Brew	24
Targa	21
SunRaster	18
SGI	15
EPS	9
FITS	6
PCX	6

Table 3.1. Results of the Advanced Imaging survey of image file formats. N.B. TIFF = Tagged Image File Format; GIF = Graphics Interchange Format; VIFF = Visualisation/Image File Format; PBM = Portable Bit-Map; EPS = Encapsulated PostScript; FITS = Flexible Image Transport System; PCX = PC graphiX.

It should be noted that all the formats mentioned here can represent both raw and compressed images. For example, the third byte of a PCX format image is set to 0 if the image is uncompressed and 1 if it is run-length encoded. The TIFF specification similarly records any compression techniques employed.

TIFF is more powerful and so more widely applicable than the others. It has massive flexibility built into it in the form of labelled fields (known as "tagged" fields) which can be added as required to create a very flexible representation. To manage the complexity that its power generates TIFF uses the notion of "classes" which split the fields into groups, forming minimum sets for certain applications. This allows TIFF writers to be fairly simple; TIFF readers, however, remain relatively complex.

The designers of the TIFF format sought to achieve three technical goals [50]:

1. The format should be extendable. It should allow new types of image to be added to its repertoire without invalidating previously created TIFF files.

2. The format should be portable. TIFF should be TIFF regardless of the hardware platform and operating system on which it is implemented.

3. The format should be revisable. It should be not only an efficient medium for exchanging image information but also be usable as an internal data format for image editing applications.

The result was a rich, but complex file structure which subsumes many previous designs. TIFF is controlled by a specification jointly written by the Aldus Corporation and Microsoft, although many other companies contributed to its formulation. The TIFF format is in the public domain, so those using the convention incur neither fees nor royalties. A detailed description of TIFF can be found in [50].

Several attempts have been made to develop a single, vendor-independent, universal image file standard, though none have been truly successful. The most recent effort in this direction began in 1990 under the auspices of the ISO. Known as the IPI (Image Processing and Interchange) standard (ISO-12087), this was intended for use in both image transfer and processing [49].

As well as gauging popularity, the Advanced Imaging review [49] provides other useful information about the various image file formats. Perhaps the most interesting evaluation criteria considered are complexity (more formally referred to as the transformation index) and hi-fi taxonomy. The complexity rating is a measure loosely related to the number of transformations that must be applied to images stored in the given format before they can be displayed. The hi-fi taxonomy asks two basic questions of the internal structure of image files:

1. Is the image stored as a simple (flat) array or are the constituent pixels grouped to form a hierarchical structure?

2. Does the structure support explicit labelling of image properties, allowing the user to add information beyond simple pixel values, or must this information be implicitly represented within the image array?

The answers to these questions assign each file format to one of four classes: HE (Hierarchical structures with Explicit labels), HI (Hierarchical structures with Implicit labels), FE (Flat structures with Explicit labels) and FI (Flat structures with Implicit labels). The use of field hierarchy and labels increases the flexibility of a format but places an overhead on any format reader that must be able to deal with these structures. It was generally found that such structures increase the complexity rating.

While popularity gives evidence of its potential user base, the hi-fi taxonomy can be used to assess the flexibility of a format and complexity (transformation indices) can be used to give an indication of the relative difficulty of writing programs to deal with it. Such information can be difficult to acquire from the varied and frequently complex technical format specifications published. Choice of format is a compromise: programming complexity, flexibility and popularity must be balanced as effectively as possible. Reviews like that discussed here appear periodically in the image processing press and serve as good starting points to those seeking a suitable image file format. The alternative is to spend significant periods of time examining rather lengthy and complex descriptions of the standards themselves.

Chapter 4
Binarisation

4.1 A Taxonomy of Thresholding Techniques

Having introduced the line drawing interpretation problem, overviewed the technology underlying current approaches to its solution and considered the acquisition of line drawing images in some detail, we now turn our attention to the first of the transformations to be applied to the image; binarisation. As noted in section 2.2, binarisation can occur at any of a number of locations in the processing chain and may be integrated with other processes. In what follows, however, we treat binarisation as a distinct, independent process. The integration of binarisation and vectorisation is considered in Chapter 7.

Binarisation, the reduction of a grey level image to a binary array, is achieved via thresholding. Let A $\{a_{ij}\}$ be a grey level image and b_0, b_1 be a pair of individual grey levels. Thresholding produces a binary image B by generating a new binary pixel b_{ij} from each pixel a_{ij} of A as follows:

$$b_{ij} = \begin{cases} b_0, & \text{if } a_{ij} <= t_{ij} \\ b_1, & \text{if } a_{ij} > t_{ij} \end{cases} \quad (4.1)$$

where t_{ij} is an appropriately selected threshold value. It is standard practice, though not essential, for b_0 and b_1 to take values corresponding to black and white in the image representation being used. The assumption underlying thresholding is that the input image may be segmented into, and therefore completely represented by, two sets of regions; one corresponding to objects, the other to background. In line drawing images, pen strokes usually constitute objects while blank paper forms the background.

Binarisation is widely used in image processing and analysis. Many thresholding schemes have been developed and described in detail in the literature; [51-53] provide good reviews of this material, with [53] providing perhaps the most in-depth technical coverage. The common issue is the identification of appropriate values of t_{ij}. Indeed, the nature and properties of the threshold(s) employed are often used to classify binarisation techniques. Figure 4.1 shows one possible taxonomy.

In global thresholding $t_{ij} = T$ for all image locations i,j; i.e. the same threshold value is applied to every pixel. Global thresholding is appropriate if the intensity values arising from objects and background are fairly consistent over the image. This would normally be true, for example, of a high quality scan of a high contrast drawing produced using a new pen and very clean, smooth paper. Any significant non-uniformity in the grey level of the paper or ink, however, reduces the effectiveness of global thresholds; it becomes increasingly difficult to select a threshold that is equally suited to all areas of the image. Non-uniformity might arise from something as simple as having a pen run out and replacing it with a new one, or having a poorly aligned lamp within the scanner unit. Some drawing media are more prone to non-uniformity than others. The problem is particularly pronounced, for example when dyeline copies, commonly used for drawing storage in mechanical engineering, must be input. The background regions of even good quality dyeline drawings are often highly variable.

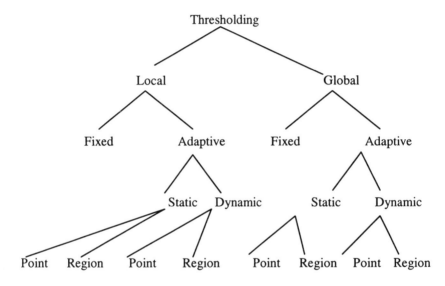

Figure 4.1. A taxonomy of thresholding techniques.

When object or background grey levels are known, or can be assumed, to vary significantly across the image a local thresholding technique is more suitable. In local thresholding the input image is divided into a number of regions, each of which is associated with a (potentially) different threshold value. These regions are often fixed; a common technique is to divide the image into a number of square or rectangular sub-images, (cf. figure 2.3). The regions used can, however, be determined dynamically on the basis of the grey level statistics of the input image. In the extreme it is possible, though highly unlikely, to allocate a different threshold value to each pixel.

A simple, if somewhat crude, way to assess the performance of a local thresholding scheme is to ask whether the boundaries of the regions used are visible in the binarised image. The advantage of local thresholding is its potential to tune thresholds to the varying local properties of the image. If this has been achieved the binarised image should accurately and only represent the separation between ink and paper in the original drawing. When there is a mismatch between the varying thresholds and the varying image content, systematic noise in the form of wrongly classified pixels will appear at or near region boundaries. Local grey level properties usually vary smoothly across the image, requiring a similar smooth variation in threshold level. Boundary noise is introduced when the threshold applied to one region differs greatly from that applied in an adjacent region. Visible boundaries suggest that either smaller regions or a different threshold selection criterion should be considered. It should be noted, however, that it is very difficult to achieve perfect binarisation with any method and that it is fairly common practice to apply some form of smoothing technique, after binarisation, to eliminate any discontinuities between the regions of a locally thresholded image [53,54].

Having determined the region (perhaps the entire image) to which a threshold is to be applied, that threshold value must be determined. This can be achieved in either of two ways. The easier to implement but often the most difficult upon which to base a working system is the fixed threshold. Here the user is asked to provide a t_{ij} for each region to be binarised. To produce a reasonable estimate the user will probably enter into an iterative, trial and error process, applying a series of thresholds until results are obtained with which he/she is satisfied. This can result in the identification of threshold values which are stable (i.e. effective) over a reasonable set of images.

The drawback, however, is that if a drawing or image type is encountered which the user did not include in his/her "training" set the threshold(s) chosen may be inappropriate. Setting the threshold value too low results in the appearance of spurious black regions in the midst of what should be considered plain paper. Pen strokes will also appear artificially thick and can display fairly wide variations in width. Application of an excessively high threshold, in contrast, misses fine detail completely and can introduce breaks in even quite thick lines. The global nature of the thresholding can be reflected in binary images whose quality varies as a function of position. Moreover, as the drawing interpretation system simply applies the user-defined value without further reference to the image there is no way the system can detect or alert the user to any problems.

The second, potentially more robust, approach is adaptive threshold selection. Here the user does not specify thresholds directly. Instead, threshold values are defined as functions of the properties of the region under consideration. As such, they are usually calculated automatically. The majority of adaptive threshold selection techniques begin with the creation of a histogram of the grey levels present in the region to be binarised. Assumptions regarding the likely shape of this histogram are made and used to specify a relationship between a suitable threshold value and some feature or features of the intensity distribution. One advantage of this approach is that tests can be made on the histograms obtained and a warning generated when the intensity distribution is not as it was expected to be. Adaptive techniques also pass responsibility for threshold selection from the user/operator to the system designer. This reduces the amount of detailed knowledge and/or experience the operator must have to work with the system and so increases the likelihood of industrial acceptance.

Weska [51] defines a threshold operator to be a test involving a function $T(x, y, N(x,y), g(x,y))$, where $g(x,y)$ specifies the grey level of the pixel at location x, y, and $N(x,y)$ is some function of the local image neighbourhood. She goes on to define adaptive thresholds to be those which depend on both $g(x, y)$ and $N(x, y)$, and finally describes threshold selection methods which vary with g(x,y), N(x,y) and the coordinate values x, y as dynamic. Since dynamic thresholding methods must by this definition be adaptive, we introduce in figure 4.1 the concept of static adaptive techniques to identify adaptive methods that are not dynamic.

In a static adaptive method the threshold value applied depends only upon the distribution of grey levels within the region under consideration; no information regarding the properties of other regions is included in the process. In a dynamic adaptive method other regions are considered. Specifically, the threshold applied to a given region is a function of the position of that region relative to other regions having particular local properties. Dynamic adaptive methods are not widely used in line drawing interpretation and so will not be considered in later sections. To make the concept clear, however, it is perhaps worth giving a brief example of a dynamic adaptive method at this point.

Chow and Kaneko [55] employed a dynamic adaptive technique in their work on radiographic images. The image was divided into overlapping, square regions measuring 7 x 7 pixels and a grey level histogram produced for each. On the basis of these histograms each region was classified as either sampling a single image entity (object or background) or spanning an object/background border. Thresholds were then computed, using a static adaptive method, for all regions deemed to contain borders. Thresholds for the other regions were subsequently determined via a linear interpolation procedure that took into account the distance of each region from the previously detected boundary areas. This latter stage is a fairly typical example of dynamic adaptive thresholding.

The final distinction recognised by our taxonomy, separating point-dependent and region-dependent threshold selection, was introduced by Sahoo et al [53]. Point-dependent methods are those which rely solely upon the grey level of each pixel: while they might involve the computation of grey level histograms and consider statistical properties of the (local) grey level distribution, the fundamental unit of data remains the image intensity measurement. In contrast, region-dependent methods also measure and exploit other image properties. A region-dependent method might, for example, compute higher-order derivatives of intensity at each pixel and base its threshold selection on the statistical properties of these derivative images [e.g. 56]. Consideration of other pointillist image properties obviously does not preclude a heavy reliance on grey level data; a common use of derivative measurements, for example, is to modify the grey level histogram in such a way as to increase the reliability of more standard point-dependent thresholding. Examples of point- and region-dependent thresholding of line drawings are considered in more detail in Section 4.3.

Throughout the preceding discussion we have implicitly assumed that only a single threshold value is required for any given image region. While this is usually true of line drawing images it is important to be aware of the possibility of multi-level thresholding. Wang and Haralick [57], for example, describe a recursive technique in which each pixel is first classified as edge (boundary) or non-edge and, on the basis of its neighbourhood, as relatively light or relatively dark. Two histograms are produced, one from edge and relatively dark pixels and one from edge and relatively light pixels. A threshold is then selected on the basis of the grey level intensity value corresponding to one of the highest peaks from the two histograms. To obtain multiple thresholds the procedure is recursively applied, first using only pixels whose intensities are smaller than the threshold and then using only those pixels whose intensities are larger than the threshold.

In the remainder of this chapter we consider the two major questions to be answered by the designer of a thresholding scheme for line drawing images; to which image regions should distinct thresholds be applied and how should those thresholds be selected? In considering these issues we implicitly focus on the most commonly employed solution category: single threshold, static adaptive techniques. Section 4.3. provides an overview of the more popular and effective approaches while sections 4.4 and 4.5 consider in greater detail the methods that we ourselves employ. First, however, it is necessary to review the document image statistics upon which binarisation methods are typically based.

4.2 Document Image Statistics

Perhaps the best-developed approach to threshold selection is to specify threshold values relative to some feature or features of a grey level histogram. Grey level histograms, as their name suggests, record the frequency of occurrence of each possible grey level over some image area. The grey level histograms arising from images of natural scenes can be complex. Single threshold binarisation methods, however, either explicitly or implicitly assume that the histograms they must deal with are bimodal. Bimodal grey level histograms comprise exactly two distinct peaks; one corresponding to objects, the other to background. In an ideal binary image each peak will be a perfect spike, in real images illumination effects, camera digitisation error and minor variations in the colours of the two viewed objects will introduce some spread.

A number of threshold selection methods may usefully be applied to images whose histograms are bimodal. One of the earliest, the P-tile method is due to Doyle [58]. Here it is assumed first that the image comprises dark objects on a light background and second that the proportion of the image occupied by objects is known. Given this information and a grey level histogram it is relatively simple to set a threshold value at the highest grey level that labels at least the expected number of pixels as belonging to an object. The P-tile approach is useful in manufacturing domains where the goal is often to locate known objects within an image, perhaps to facilitate visual inspection. If the object size and approximate viewing distance are both given, reasonable estimates of the expected percentage of object pixels can be derived. In line drawing interpretation, however, this is unlikely to be possible.

Another, more widely applicable technique is the mode method of Prewitt and Mendelsohn [59]. The basic idea behind this is to locate the minimum that must lie between the two maxima of a bimodal histogram. The grey level value at which the minimum appears is adopted as a threshold value, this being arguably the best place at which to separate the two peaks. As grey level images are often subject to noise, reliable location of peaks and troughs in their histograms can be problematic. A number of variations on the mode method theme have therefore been proposed, some of which are discussed by Rosenfeld and Kak [60]. More recently Rosenfeld and De La Torre [61], for example, have fitted convex hulls to grey level histograms and defined the threshold value to be the point at which the distance between the histogram and its hull is maximal.

The techniques outlined above all rely upon the assumption that the grey level histogram comprises two <u>clearly defined</u> peaks. Unfortunately, poor contrast and noisy images often render peak identification difficult. This can be a problem given almost any image type; images of line drawings, however, pose additional problems.

For most graphics images, the peak representing the object is very much smaller than the peak corresponding to the background intensity. Although we might perceive the drawing as spanning the paper background, the percentage of that background which is actually obscured by ink is usually very small. Figure 4.2 shows typical examples. The peak generated by the linework is barely visible. Note also that the area between the two peaks is quite flat; identification of a reliable local minimum within this region is bound to be prone to error. It could be argued that as separation of ink and plain paper would be achieved by thresholding at any value within the central plateau, each possible threshold is as good as all the others. This is not generally the case. While all the thresholds in this region are likely to produce a reasonable rendition of the original pen strokes, the level of detail visible in and available from the binarised image will vary considerably. Subsequent processes charged with the detection of small-scale drawing features, for example arrowheads, might produce widely varying results as different threshold values are applied.

Because reliable peak and valley detection is a common problem in image processing and analysis, a number of (region-dependent) procedures have been put forward which seek to modify the grey-level histogram so as to alleviate the problem. These are discussed in Section 4.3.

As the above implies, most threshold selection techniques rely on comparatively simple image statistics; often only means, variances [63] and standard deviations [64]. There are, as might be expected, some exceptions to this overly simple rule. Ahuja and Rosenfeld [65], for example, base threshold selection on the co-occurrence matrix introduced by Haralick et al [56]. The entries in Ahuja and Rosenfeld's co-occurrence matrix are measures of the frequency with which each possible pair of grey levels occurs either horizontally or vertically adjacent to each other. Pixels towards the interior of object or background regions should contribute to near diagonal elements of this matrix while boundary pixels may be expected to be significantly off-diagonal. This information may be used to create grey level histograms in which the bi-modal structure of the image is emphasised. Deravi and Pal [66] take a similar approach based on transition matrices while Kirby and Rosenfeld [67] employ scatter plots, again in similar fashion. For the most part, however, a basic understanding of grey level histograms and appropriate use of the a priori knowledge that the image to be binarised depicts a line drawing is more important than detailed consideration of complex statistical measures.

Binarisation 63

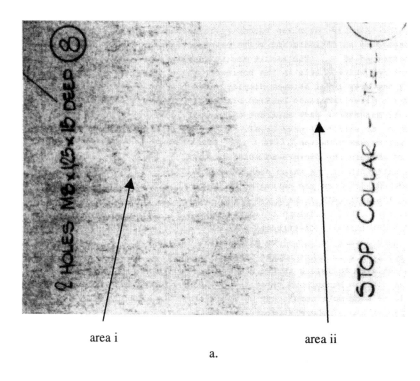

area i   area ii
a.

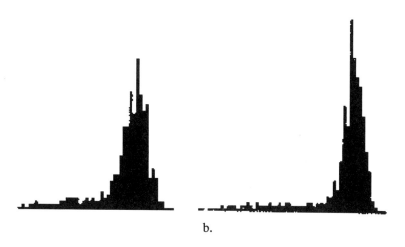

b.

Figure 4.2. a) an image of a real engineering drawing, b) grey level histograms of areas i and ii. Figures 1 and 2 of Dunn and Joseph [64], reproduced by permission.

4.3 Binarising Line Drawings

As the great majority of paper documents are bilevel in nature, it is hardly surprising that most document scanners incorporate some form of binarisation procedure. The methods employed are, however, usually quite simple (often fixed) global techniques intended only to perform well when provided with high-quality original documents. If the scanner illumination, paper reflectivity or ink colour is non-uniform, photocopying or age has reduced the document quality or significant additive noise is introduced by the electronics, on-board thresholding is unlikely to produce high quality results. A further potential drawback is that scanner manufacturers rarely publish technical details of the thresholding methods they employ; most of their customers would not be interested. Those wishing to apply further image interpretation techniques, however, often need to know what assumptions have been made regarding the input drawing and what has been done to produce the data with which they must work. Without such knowledge it can be difficult to ensure that subsequent processes are based on reasonable assumptions regarding the nature and properties of the binarised image. It may therefore become necessary, even when on-board binarisation appears satisfactory, to use the scanner's grey level image option and apply an application-specific thresholding method independently. In the remainder of this section we consider in some detail a number of binarisation techniques which were either specifically designed for use with line drawing images or which have been found to be particularly effective in drawing interpretation.

4.3.1. Point-Dependent Threshold Selection

One of the best known, point-dependent thresholding methods is the Minimum Error technique due to Kittler and Illingworth [68]. Although a multi-level version of minimum error thresholding is described [68], in its basic form the algorithm produces a single threshold value.

Minimum error thresholding assumes that the grey-level histogram is a reasonable estimate of the probability density function $p(g)$ of a mixture population containing object and background pixels. It is further assumed that the two components of the mixed population $p(g \mid i)$, where $i = 1, 2$, are normally distributed with means μ_i, standard deviations σ_i and a priori probabilities P_i. The aim of the algorithm is to identify the threshold value t for which

$$P_1.p(g \mid 1) < P_2.p(g \mid 2) \quad \text{if } g <= t$$

and (4.2)

$$P_1.p(g \mid 1) > P_2.p(g \mid 2) \quad \text{if } g > t$$

This value is the Bayes minimum error threshold; thresholding the original image at t will minimise the number of incorrectly classified pixels.

Determination of the minimum error threshold is comparatively simple. Taking the logarithm of both sides of equation (4.2) and substituting the standard equation for a Gaussian for each $p(g \mid i)$ gives a quadratic in t whose coefficients are simple

Binarisation

functions of μ_i, σ_i and P_i. If the parameters of the two normal distributions are known or can be estimated this equation can be solved to provide a threshold value. Nagawa and Rosenfeld [69] previously proposed an algorithm that involves fitting Gaussian distributions to the histogram to determine μ_i, σ_i and P_i. Kittler and Illingworth, however, reject that approach as computationally expensive and go on to propose a simpler, iterative method. In their algorithm an initial threshold value $t = T$ is first chosen arbitrarily. The parameters μ_i, σ_i P_i are then estimated for each of the two resulting sections of the distribution without explicit fitting of Gaussian profiles. These two models may be used to estimate the conditional probability $e(g, T)$ of grey level g being correctly classified after thresholding at T. $e(g, T)$ then forms the basis of an index of classification performance which reflects the amount of overlap between the two (assumed Gaussian) populations obtained by splitting the histogram at T (figure 4.3). This measure is smooth and easy to compute, making the search for a minimum value straightforward. Full mathematical details are given in the original paper; the proposed algorithm, however, is as follows [68]:

1. Choose an initial threshold T.

2. Compute $\mu_i(T)$, $\sigma_i(T)$, $P_i(T)$, for $i = 1, 2$.

3. Compute the updated threshold by solving quadratic equation (4.3) for g

$$g^2 . [1/\sigma_1^2(T) - 1/\sigma_2^2(T)] - 2.g.[\mu_1(T)/\sigma_1^2(T) - \mu_2(T)/\sigma_2^2(T)]$$

$$+ \mu_1^2(T)/\sigma_1^2(T) - \mu_2^2(T)/\sigma_2^2(T) + 2.[\log\sigma_1(T) - \log\sigma_2(T)] \quad (4.3)$$

$$- 2.[\log P_1(T) - \log P_2(T)] = 0.$$

4. If the new and old thresholds are sufficiently similar, terminate the algorithm, otherwise set $T = g$ and go to step 2.

The above algorithm is simple to implement and provides a good approximation of the minimum error threshold very quickly. Note that the method cannot provide an exact solution, as the use of a threshold to separate the two modes of the histogram will of necessity truncate both components to some degree, causing errors in their estimated parameters. Further problems can arise in that, depending upon the initial value of T, the algorithm may converge to a solution at one of the boundaries of the grey-level range. This is, however, easy to detect. Imaginary solutions to equation (4.3) may also arise. Kittler and Illingworth [68] suggest that the algorithm be run with a number of different initial values and the results be compared to ensure that its output is reliable.

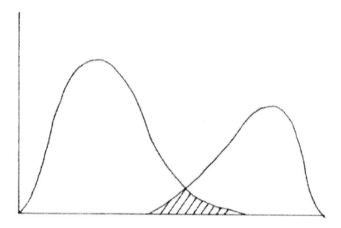

Figure 4.3. Sketch of a bimodal histogram with the overlapping area shaded.

Kittler and Illingworth have applied minimum error thresholding both globally and locally (using fixed sub-images 32 x 32 pixels square) to a variety of images. While they were more concerned with processing images of mechanical parts than line drawings, they found their method to be particularly appropriate when the two populations making up the grey level histogram are of unequal size (1:100 or more), as is usually the case with drawing images. Fan and Xie [70] have recently provided an alternative explanation of minimum error thresholding, expressed in terms of relative entropy.

While minimum error thresholding has been found effective when one peak dominates an otherwise bimodal histogram, Dunn and Joseph [64] propose a method, tuned to poor quality line drawing images, which explicitly assumes that the histogram peaks corresponding to paper and ink are of radically different sizes. Indeed, they take the view that although the peak generated by plain paper is usually well defined and amenable to modelling, the one generated by ink marks is typically so small that it cannot be reliably located. A similar observation was made by Bartz [71], whose system produces a threshold that is independent of black level whenever the darker peak in the histogram is weak. Dunn and Joseph's method, however, is always based upon simple measurements of the location and width of the white peak alone; black areas effectively being defined as those with a low probability of being white.

The proposed method is a local one. Dunn and Joseph first compute grey level histograms over 64 x 64 pixel sub-images, then smooth those histograms slightly by averaging over three grey levels. That is, the value at each histogram location becomes the mean of itself plus the values of its two neighbours. Some smoothing is necessary to allow the peak corresponding to clean paper to be accurately identified. After smoothing, Dunn and Joseph report experiments that suggest that the peak can be located to +/- one grey level, by simply seeking the modal grey level g_m. Once g_m has been identified, the noise level associated with white paper is estimated by measuring the half-width of the white peak at half its height (figure 4.4). This value is approximately 1.2σ for a normally distributed peak of standard deviation σ.

Binarisation

Experiments have shown [64] that although the half-width may vary somewhat from sub-image to sub-image this variation does not seem to relate to visually obvious variations in background noise. The mean half-width across all sub-images was therefore used as a background noise measure. This global measure was, however, used in conjunction with each g_m to generate local threshold values; thresholds were set at g_m - 3 average half-widths, i.e. g_m - 3.6σ. Figure 4.5 shows the result of applying this procedure to a dyeline copy of an engineering drawing.

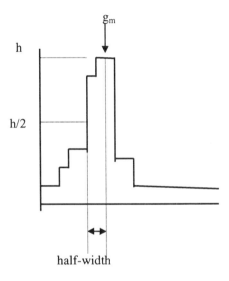

Figure 4.4. Dunn and Joseph's measure of noise in the grey level histogram of clean paper.

Dunn and Joseph's method has several strengths. First, making the threshold a function of the standard deviation of the white peak means that, assuming a normal distribution of white pixels, the noise level expected from a given threshold may be estimated. The method also relies only upon the most reliable section (the top half) of the most reliable histogram peak. Comprehensive experimental evaluation [64] also shows the technique to deal well with the systematic noise that so often arises in dyeline and electrostatic copies of line drawings.

Although Dunn and Joseph's method could be used to binarise an input image, and the results it produces are impressive, their intention is that the information provided by their method be used instead to normalise the grey level data. Normalisation produces an image in which the mean white level is zero and darker regions are increasingly positive. Later processes, they suggest, should have access both to the statistical measures they compute and the full grey level image. We shall return to this idea again in section 4.5 and Chapter 11.

Figure 4.5. Application of Dunn and Joseph's method to the image of figure 4.2.a. Figure 4 of Dunn and Joseph [64], reproduced by permission.

One further point-dependent method is particularly relevant to line drawing interpretation. O'Gorman and Kasturi [72] report that in their extensive work on document image analysis the moment preserving thresholding technique of Tsai [73] has consistently produced impressive results. Tsai's approach is to consider the input image to be a blurred version of a true binary image and to attempt to remove the blur. This is done by computing several moments of the image before thresholding and choosing a threshold such that those moments remain unchanged in the binarised image. The ith moment m_i of an image is defined by:

$$m_i = (1/n) \sum_x \sum_y f(x, y) \qquad i = 1, 2, ... \qquad (4.4)$$

where $f(x, y)$ specifies the grey level at location (x, y) within an image f comprising n pixels. Tsai, who is only concerned with $i = 1,2,3$, proposes a deterministic method which selects thresholds without iteration or search. The method may also be extended to multi-level thresholding; details of moment preserving bi-level, tri-level and quadri-level thresholding are given in [73]. For the present we are only concerned with the bi-level algorithm.

As functions of the image's grey level distribution, rather than its geometric properties, moments may be calculated either directly from a raster array or from a grey level histogram. In the latter case, m_i is given by:

$$m_i = (1/n) \sum_j n_j (z_j)^i = \sum p_j (z_j)^i \qquad (4.5)$$

where n_j is the number of pixels with grey level z_j and $p_j = n_j/n$. The aim of moment preserving thresholding is to identify a threshold value such that replacing all sub-

Binarisation

threshold grey levels by z_0 and all supra-threshold grey levels by z_1 results in a binary image g whose moments m_i' for $i = 1,2,3$ are equal to the corresponding moments of f. Let p_0 and p_1 be the fractions of sub- and supra-threshold pixels in f respectively, then:

$$m_i' = \sum_{j=0}^{1} p_j (z_j)^i \qquad (4.6)$$

$$p_0 + p_1 = 1 \qquad (4.7)$$

and the requirement that $m_i' = m_i$ for $i = 1,2,3$ produces a set of four equations

$$\begin{aligned} p_0.z_0^0 + p_1.z_1^0 &= m_0 \\ p_0.z_0^1 + p_1.z_1^1 &= m_1 \\ p_0.z_0^2 + p_1.z_1^2 &= m_2 \\ p_0.z_0^3 + p_1.z_1^3 &= m_3 \end{aligned} \qquad (4.8)$$

which may be solved to find p_0, p_1, z_0 and z_1. Once p_0, p_1 have been obtained, the P-tile method of [58] (see section 4.2) may be applied. Note that z_0 and z_1 represent estimates of the grey level of ink marks and plain paper respectively. Equations (4.8) may be solved as follows [73]. First let

$$c_d = \begin{vmatrix} m_0 & m_1 \\ m_2 & m_3 \end{vmatrix}$$

$$c_0 = (1/c_d). \begin{vmatrix} -m_0 & -m_1 \\ -m_2 & -m_3 \end{vmatrix}$$

$$c_1 = (1/c_d). \begin{vmatrix} m_0 & -m_1 \\ m_2 & -m_3 \end{vmatrix}$$

then

$$z_0 = (1/2).[-c_1 - (c_1^2 - 4.c_0)^{1/2}]$$

$$z_1 = (1/2).[-c_1 + (c_1^2 - 4.c_0)^{1/2}]$$

Finally,

$$p_d = \begin{vmatrix} 1 & 1 \\ z_0 & z_1 \end{vmatrix}$$

$$p_0 = (1/p_d) . \begin{vmatrix} 1 & 1 \\ m_1 & z_1 \end{vmatrix}$$

and

$$p_1 = 1 - p_0.$$

Tsai [73] shows several examples of the operation of this method on a variety of grey-level images, some of which depict (textual) documents. Figure 4.6 shows the result of applying the method to an image containing English and Chinese text.

a.

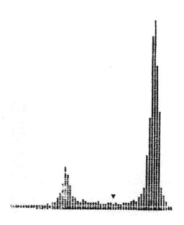
b.

c.

Figure 4.6. Moment preserving thresholding; a) the input image, b) the grey level histogram of the image with the chosen threshold value marked by a triangular pointer, c) the result of applying the threshold to the original image. Figure 2 of Tsai [73], reproduced by permission.

It is hard to give a clear recommendation as to which of the algorithms described here is best suited to images of line drawings. The method of Dunn and Joseph was specifically designed for poor quality engineering drawings. Those due to Kittler and Illingworth and Tsai are general techniques that have been found effective given line drawing data. The performance of any thresholding algorithm depends upon both the shape of the histograms obtained from the images to hand and the goal to be achieved. All that can really be said is that, when seeking a thresholding method for a given line drawing interpretation task, each of the above techniques is worthy of consideration and experimentation.

4.3.2 Region-Dependent Threshold Selection

A common theme underlying region dependent threshold selection methods is histogram modification. Two approaches can be identified. The first seeks to create well-defined bimodal histograms so that methods which seek a plateau separating two peaks [59-61] may be applied with a reasonable chance of success. The second attempts to produce a uni-modal histogram in which the modal grey level is an appropriate threshold value.

Regardless of the approach taken, the general solution to the histogram modification problem (Weska and Rosenfeld [74]) is to create a two-dimensional scatter plot in which one axis specifies grey level and the other some measure of the likelihood that a given pixel lies on an object/background boundary. A modified grey level histogram may then be produced by weighted projection of the scatter data onto the grey level axis. That is, the value each pixel contributes to the grey level histogram is not fixed, as is usually the case; instead it reflects in some way the likelihood that that pixel lies on a boundary. Emphasising, i.e. giving a higher weight to, pixels which are unlikely to lie on boundaries should produce an improved bimodal histogram. Increasing the contribution made by suspected boundary pixels, on the other hand, should produce a single peak centred on a value intermediate between the object and background grey levels. Thresholding at the mode or mean of this peak ought to lead to effective binarisation.

The need to detect, or at least highlight, boundaries between regions of more or less constant grey level is common in image analysis and a multitude of boundary or edge detection algorithms therefore exists (Ballard and Brown[32], Hlavac et al [34], Trucco and Verri [35]). The majority assume that the boundaries they seek can be thought of as step changes in grey level that may be identified by examining the first and/or second derivatives of image intensity. Step changes generate extremes in the first derivative of grey level and zero-crossings in the second, both of which may be detected automatically without undue effort.

The widespread use of derivative estimates in edge detection has led to a similar reliance on such measures in the formation of the scatter plots from which modified grey level histograms are generated. Perhaps the best known histogram modification method is that developed by Mason et al [62] in their work on images of human chromosomes. In Mason et al's technique the contribution made by a given pixel is inversely proportional to a measure of the first derivative of intensity made across its local neighbourhood. If Δ is an estimate of the first derivative of intensity at a given pixel, that pixel's contribution to the grey level histogram is given by

$$\frac{1}{(1+|\Delta|^2)} \qquad (4.9)$$

This gives full weight (1.0) to pixel values that typically appear in smooth regions of the image and negligible weights to those which tend to appear near the sharp changes in image intensity which mark region boundaries. The overall effect is to sharpen peaks in the grey level histogram which correspond to object and background, making it easier to site a threshold value between them. A similar method is advocated by Watanabe [75], while Katz [76] takes the complementary approach and describes an algorithm in which weights are directly proportional to local estimates of the first derivative.

Weska, Nagel and Rosenfeld [77] suggest use of the second derivative, rather than the first, to emphasise the natural bimodality of the grey level histogram. They note that while the second derivative is zero on the linear ramp of a step edge it has high values on either side. A weighting function based on the second derivative (in [77] the Laplacian is employed) can therefore enhance both background and object grey levels while suppressing boundary pixels.

While histogram modification is most commonly based upon derivative measures, other approaches are possible. Wu et al [78], for example, propose the use of quadtrees to provide grey-level weighting functions. A quadtree is formed by recursively sub-dividing a grey level image into four equal quadrants until each quadrant satisfies some test of grey level homogeneity. Wu et al continue the division process until the standard deviation of grey level within each (sub-)quadrant falls below a threshold. They argue that once this process is complete, smaller quadrants are more likely to be close to boundaries than larger ones and that the size of the (sub-)quadrant within which a given pixel lies may therefore be used to determine its contribution to a modified grey level histogram.

Histogram modification techniques have been both concisely reviewed and experimentally evaluated by Weska and Rosenfeld [74]. While these techniques were found to be effective given a fairly wide variety of images, their performance on images of linework was disappointing. Moreover, the reasons for this poor performance were quite fundamental and suggest that region-dependent threshold selection for line drawing images might be problematic in principle.

A key feature of line drawing images is that the majority of the object (linework) regions have lengths that are many times their width. As a result, nearly all linework pixels are also boundary pixels. In Weska and Rosenfeld's experiments this caused methods which emphasise low first derivative values to fail; almost no zero (or very low) derivative values were obtained from object pixels, so there was no discernible black peak in the modified histogram. That line drawing images typically contain many more background than object pixels only serves to emphasise this effect. Similarly, Weska and Rosenfeld also found that giving high weights to boundary pixels often fails to produce uni-modal histograms from line drawing images. Better results were obtained from the Laplacian method of Weska, Nagel and Rosenfeld

Binarisation 73

[77], in that clearly bimodal histograms were generated. Closer examination, however, revealed that many of the contributing pixels did not lie on a boundary but were simply isolated points of extreme grey level; i.e. noise. This clearly casts doubt upon the reliability of the method.

In essence, all the above techniques assume that the input image comprises three sets of pixels; object, background and boundary, each of which is large enough and different enough from the others that it may be identified. In line drawing images this is often not true. While these methods may be applicable given high resolution images of the more compact drawing entities such as characters and symbols, they are unlikely to be effective on even average quality images of linework.

A valuable region-dependent post-process, proposed by Yanowitz and Bruckstein [79], is, however, reported by Trier and Taxt [80]. This centres upon computation of the average intensity gradient around the boundary of each black region produced by binarisation. Objects having an average gradient lower than some threshold value T_p are considered to have been misclassified and are subsequently removed (i.e. turned white). The method, which enforces a requirement for high contrast edges between paper and ink, is as follows [80]:

1. smooth the original image by a 3 x 3 mean filter to reduce noise;

2. estimate the gradient magnitude of the smoothed image using the Sobel edge detection operator (see e.g.[32]);

3. manually select a threshold value T_p;

4. for each black region, calculate the average gradient over the set of member pixels that are adjacent to at least one non-member (i.e. background) pixel;

5. if the computed average for any region is less than T_p, remove that region.

Trier and Taxt [80] describe a series of experiments aimed at evaluation of a range of previously published line drawing thresholding methods. They initially found the method of Yanowitz and Bruckstein [79] to give the most satisfactory overall performance, but later discovered that incorporation of Yanowitz and Bruckstein's post-processing stage allowed several other techniques to overtake the original. It seems likely, therefore, that the algorithm described above might usefully be employed in conjunction with any line drawing binarisation approach. Note that this algorithm uses region-dependent information to make a <u>local</u> judgement as to the likelihood of a boundary being present. This is in contrast to the methods described previously, which attempt to incorporate region-dependent data into more global statistical measures where it is overwhelmed.

4.3.3 Image Partitioning

The techniques discussed above may be applied either globally or locally to the input image. In those cases where local thresholding is considered desirable, a key question is to which image sub-region(s) should the selected method be applied? On the one hand, each region should be of sufficient size to ensure that the number of background pixels included is large enough to allow a reasonable estimate of background noise to be obtained. If the sub-images used are too small, patterns of linework may not be distinguishable from short-range variations in background noise. The effect of any noise removal operations carried out on small regions may therefore be to delete, or at least damage, sections of significant linework. Dunn and Joseph [64] also point out that to locate the mean of a grey level histogram to within half a grey level to 95% confidence requires about 150 data points from a normally distributed population; hence their own use of 64 x 64 pixel regions. In contrast, however, sub-images must not be so large that they compute grey level averages over areas of significantly non-uniform noise. Ideally, the sub-regions used in a local thresholding scheme should reflect areas within which image noise is more or less constant. The obvious problem is that noise varies from drawing to drawing and image to image and so, while fixed regions are commonly used, it is difficult to guarantee that any fixed pattern of sub-images will be appropriate for any given image.

One solution to this problem is to use domain-dependent information to ensure that the results of binarisation give the expected features. To do this is to integrate binarisation into the higher-level interpretation processes, Joseph [81] has proposed a direct vectorisation method which combines binarisation and vectorisation and is described in some detail in Chapter 7. Joseph and Pridmore later integrated both these operations into a knowledge-directed entity extraction system applicable to mechanical piece-part drawings (Chapter 11).

No matter how well developed the thresholding method(s) employed, one can never expect perfect results. Though the effect varies with the quality of the original image, binarisation typically introduces distortions such as breaks in linework, ragged edges on region boundaries and spurious areas of black pixels. Similar comments could be made about each of the interpretation steps outlined in Chapter 2. The standard response to this problem is to perform each individual step as effectively as possible but also to defer decisions that do not need to be made until later. This approach, termed the Principle of Least Commitment by Marr [33] helps avoid irreparable errors. Later in the processing sequence more and higher-level information is usually available, providing greater context and simplifying the decision making process. In some cases it may even be that difficult decisions identified early in the interpretation may not have to be made at all; the image regions in which the data concerned lies might, for example, turn out not to be part of the drawing.

Chapter 5
Binary Image Processing and the Raster to Vector Transformation

5.1 Raster to Vector Conversion

Document scanning and binarisation technologies are now quite well developed. As a result, many line drawing interpretation systems and projects take as their starting point a binary image of the original drawing. As outlined in figure 2.1, the interpretation of this image typically involves vectorisation, entity extraction, two-dimensional scene formation and, possibly, 3D reconstruction (depending upon drawing type). In this and the following two chapters we consider the steps involved in extracting a vector representation from the binary image. This process is known as the raster to vector transformation [46, 82-85].

The raster to vector transformation typically comprises a number of distinct steps (figure 5.1). While scanning and binarisation systems can be impressive, the initial binary image is unlikely to be perfect. Hence the first stage in the raster to vector conversion process is usually an attempt to remove, or at least minimise, noise. The input and cleaned images may be represented as full 2D arrays, as discussed in Chapter 2, or compressed using run-length coding or one of the other techniques outlined in Chapter 3.

If binarisation and noise removal are successful, each line in the original drawing will generate an elongated region of black pixels in the binary image. The next step is to reduce each such region to a one-pixel wide string of pixel locations: this may be achieved either by finding the centre line of each region (skeletonisation) or extracting its bounding contour (for contouring see Chapter 6). This stage is necessary as the final step, vectorisation, usually requires thin data in the form of pixel strings. The aim of vectorisation (Chapter 7) is to fit lines and curves to the input ink marks, a task which can be both slow and error prone if the ink mark is represented by a wide swath of pixels. The vectorisation process will usually associate other properties of the line, most commonly its width, with its geometrical description and may also report features such as corners and junctions. The result of the raster to vector transformation is an intermediate database from which the recognition of image constructs, i.e. engineering drawing entities and cartographic objects, may begin.

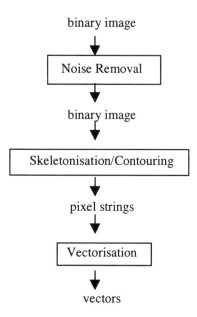

Figure 5.1. The raster to vector transformation.

Some confusion can arise from the use of the terms "raster to vector transformation" and "vectorisation". In what follows we shall use the former to denote the entire process of noise removal, skeletonisation/contouring and the recovery of a vector representation. We shall reserve the term "vectorisation" for the operation that extracts a vector description from thin line data. In much of the literature, however, "vectorisation" is used to refer to the entire raster to vector transformation.

In the remainder of this chapter we consider the first stage in the raster to vector conversion process: the removal of noise from a binarised image. To do this it is necessary first to introduce some of the basic terminology and concepts of binary image processing. In doing this we do not aim to give a comprehensive overview of that field; many other books are devoted to the area. Instead, we merely aim to introduce terms and ideas which will be used in this and subsequent chapters. The reader may choose to consider these sections in the order in which they appear or to return to them when an understanding of the concepts they introduce is required. Section 5.2 gives some basic nomenclature and definitions. Sections 5.3 and 5.4 introduce distance transforms and mathematical morphology respectively. Consideration of general and line-drawing-specific noise reduction algorithms begins in section 5.5.

5.2 Some Definitions

Let $A = \{a_{ij}\}$ be a binary image digitised on a rectangular grid. Let B and W be respectively the sets of black (object) and white (background) pixels which constitute that image. Each pixel has eight neighbours, numbered according to the following scheme (figure 5.2):

a_4	a_3	a_2
a_5	a_{ij}	a_1
a_6	a_7	a_8

Figure 5.2. The 8-neighbourhood of a pixel.

$S_8(a_{ij})$ is the set of all neighbours of a_{ij} (excluding a_{ij} itself) and is termed the 8-neighbourhood of a_{ij}. Neighbours with odd indices are direct or 4-neighbours of a_{ij} and form the set $S_4(a_{ij})$; neighbours having even indices are indirect neighbours of a_{ij} and form the set $S_D(a_{ij})$. The general term "neighbourhood" refers to S_8. The neighbour relation defines a topology on the digital plane [86].

Any pixel in B having all its neighbours also in B is said to be an interior pixel. The set of all interior pixels of B is known as the kernel or interior of B; those pixels in B that are not interior are termed contour pixels (figure 5.3).

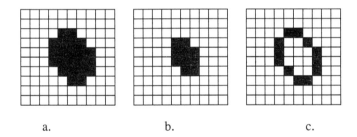

Figure 5.3. a) A simple binary image, b) its kernel and c) its contour.

One of the more important concepts in binary image processing is the notion of distance within the digital plane. We shall use the definition given in [87]: the distance between two pixels, X and Y, is the length of the shortest path connecting X and Y in an appropriate graph. It can be shown [88] that any distance may be defined by choosing an appropriate neighbourhood relation and definition of path length.

A commonly used example of a distance measure is the Euclidean distance, defined thus:

$$D[a_{i1j1}, a_{i2j2}] = [(i_1 - i_2)^2 + (j_1 - j_2)^2]^{1/2} \qquad (5.1)$$

This formula provides precise, real-valued distance estimates. The use of non-integer distance measures can, however, cause problems. As a result, various integer approximations to the Euclidean distance are in use. These may be summarised as follows:

$$D[a_{i_1 j_1}, a_{i_2 j_2}] = \begin{cases} w_0 |i_1 - i_2| + w_1 |j_1 - j_2|, & \text{if } |i_1 - i_2| > |j_1 - j_2| \\ w_1 |i_1 - i_2| + w_0 |j_1 - j_2|, & \text{otherwise} \end{cases} \quad (5.2)$$

where $w_0, w_1 > 0$.

A variety of measures may be obtained by manipulating w_0 and w_1. For example, if $(w_0, w_1) = (1,1)$ the so-called city block, or 4-neighbour, distance is measured. If $(w_0, w_1) = (1,0)$ the chessboard, or 8-neighbour, distance is reported. As one might expect, 4-neighbour distance measures the length of the shortest path in which adjacent pairs of pixels are 4-neighbours. 8-neighbour distance measures the shortest path in which adjacent pixels are 8-neighbours.

Each distance measure has its own characteristic shape, defined by the set of all pixels equidistant from some central pixel. Hence the characteristic shape of Euclidean distance is a circle. The characteristic shapes of the city block and chessboard measures are the diamond and square respectively. Figure 5.4 shows examples of the application of these distance measures to a trivial binary image.

```
0 0 0 0 0 0 0           3.6 2.8 2.2 2.0 2.2 2.8 3.6
0 0 0 0 0 0 0           3.2 2.2 1.4 1.0 1.4 2.2 3.2
0 0 0 x 0 0 0           3.0 2.0 1.0  x  1.0 2.0 3.0
0 0 0 0 0 0 0           3.2 2.2 1.4 1.0 1.4 2.2 3.2
0 0 0 0 0 0 0           3.6 2.8 2.2 2.0 2.2 2.8 3.6

        a.                          b.

5 4 3 2 3 4 5           3 2 2 2 2 2 3
4 3 2 1 2 3 4           3 2 1 1 1 2 3
3 2 1 x 1 2 3           3 2 1 x 1 2 3
4 3 2 1 2 3 4           3 2 1 1 1 2 3
5 4 3 2 3 4 5           3 2 2 2 2 2 3

        c.                          d.
```

Figure 5.4. a) A simple image with, b) Euclidean, c) city block and d) chessboard distances.

Each distance measure defines a metric. A metric on a set X is a function

$$D: X \times X \rightarrow R^+$$

which is

- positive definite: $D(x,y) = 0$ if and only if $x = y$,
- symmetric: $D(x,y) = D(y,x)$ for all x and y,
- triangular: $D(x,z) < D(x,y) + D(y,z)$ for all x, y and z.

The 8-metric based on chessboard and the 4-metric based on city block distances are perhaps the most commonly used in the processing of binary images.

Patterns of connectivity within an image often provide important clues to its interpretation. Two pixels (both members of either B or W) are said to be connected if they are neighbours (i.e. have a distance between them equal to 1) in the chosen distance metric. A connected component of an image is a set of connected pixels. To give an example relevant to line drawing interpretation, depending upon the selected metric the set of pixels comprising a line segment may be 8-connected, 4-connected or exhibit mixed connectivity (figure 5.5).

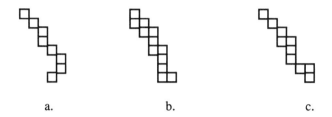

Figure 5.5. a) 8-connected, b) 4-connected and c) mixed connectivity lines.

It is common practice to apply 4- and 8-connectivity to W and B, respectively. This is to avoid the contradictions that can arise when either form of connectivity is applied to both pixel sets. Consider the example shown in figure 5.6. This clearly shows a connected object with a single hole against a separate background. Suppose we apply 8-connectivity to both B and W. We have an 8-connected object, but also a single 8-connected white region; the hole and background cannot be distinguished. If, however, we use 4-connectivity for both object and background both are unconnected. These problems can be neatly avoided, if one type of connectivity is used to define connected components in the background and the other to identify objects.

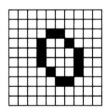

Figure 5.6. An 8-connected object with a 4-connected background.

The basic notions of distance and connectivity are typically used to define and identify structural properties and features of the binary image. To this end two additional measures, based on connectivity, are commonly used; the connectivity and crossing numbers. To define them, we introduce the following pseudo-Boolean local functions of A [89]:

$$A_4(a_{ij}) = \sum_{k=1}^{4} a_{2k-1} \qquad A_8(a_{ij}) = \sum_{k=1}^{8} a_k$$

$$B_8(a_{ij}) = \sum_{k=1}^{8} a_k a_{k-1} \qquad C_8(a_{ij}) = \sum_{k=1}^{4} a_{2k-1} \overline{a_{2k}} \, a_{2k+1} \qquad (5.3)$$

$$N_{C_4}(a_{ij}) = A_4(a_{ij}) - C_8(a_{ij}) \qquad N_{C_8}(a_{ij}) = \overline{A_8}(a_{ij}) - \overline{C_8}(a_{ij})$$

$$N_{cm}(a_{ij}) = A_8(a_{ij}) - B_8(a_{ij}) + C_8(a_{ij})$$

$$C_n(a_{ij}) = A_8(a_{ij}) - B_8(a_{ij})$$

where $a_9 = a_1$, $\overline{A_4}(a_{ij}) = \sum_{k=1}^{4} \overline{a_{2k-1}}$, $\overline{a_k} = 1 - a_k$

The values N_{c4}, N_{c8} and N_{cm} are the connectivity numbers for 4-, 8- and mixed connectivity respectively and show the number of connected components of B present in S_8. C_n is known as the crossing number [90] and shows the number of transitions from 1 to 0 in S_8 when the eight pixels are visited counter-clockwise. The connectivity and crossing numbers can be used directly to identify significant features of a binarised line drawing. Given a pixel a_{ij} in B:

a_{ij} is
- an isolated point if $\{A_4, N_c, C_n\} = 0$
- an end point if $\{A_4, N_c, C_n\} = 1$ (5.4)
- a connectivity point if $\{N_c, C_n\} = 2$
- a branch point (or node) if $\{N_c, C_n\} = 3$

Examples of these features are shown in figure 5.7. It should be noted that the point classification illustrated in figure 5.7 relies upon the image only containing lines which are one pixel wide; scanning and binarisation of real line drawings is unlikely to produce such an image. Some further processing is required (Chapter 6).

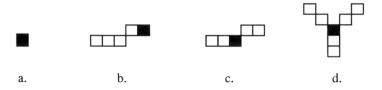

a. b. c. d.

Figure 5.7. Examples of a) the isolated point, b) the end point, c) the connectivity point and d) the node.

5.3 The Distance Transform

Most algorithms applied to binary images rely upon image parameters that are functions of distance. To compute the distance between two arbitrary pixels is a global operation: one must have access to the entire image to be certain of identifying the most appropriate path between the pixels of interest. It has been shown, however, that approximations to any desired global distance can be computed using only local operations [91]. This observation led to the creation of the distance transform, one of the most widely used tools in binary image processing.

The distance transform estimates the distance between each pixel in one set, B say, to the nearest pixel in some other set, e.g. W. The output grey level image is known as a distance map. This is a replica of B in which each pixel is labelled with its distance from W. The distance map can be computed for both B and W. In this case, W pixels are usually assigned negative distance values (figure 5.8). The fundamentals of the distance transform have been considered in some detail [92,93] and a number of specific distance transforms have been proposed and examined [94-98].

-3	-2	-2	-2	-2	-2	-2	-3	-3
-2	-2	-1	-1	-1	-1	-2	-2	-3
-2	-1	-1	1	1	-1	-1	-2	-2
-2	-1	1	1	1	1	-1	-1	-2
-2	-1	1	2	2	1	1	-1	-2
-2	-1	1	1	2	1	1	-1	-2
-2	-1	-1	1	1	1	-1	-1	-2
-2	-2	-1	-1	1	-1	-1	-2	-2
-3	-2	-2	-1	-1	-1	-2	-2	-3
-3	-3	-2	-2	-2	-2	-2	-3	-3

Figure 5.8. A distance map computed using chessboard distance for both B (positive) and W (negative) pixels.

The values recorded in distance maps are path-based; horizontal, vertical and diagonal unit moves are permitted and weighted according to the distance metric used. Weights are stored in two-dimensional arrays, or masks, and the distance transform is computed by repeated application of these masks to the developing distance map. Each mask application propagates information. Once distance has been computed for a given pixel a_{ij}, the mask determines the relation between this value and those assigned to a_{ij}'s neighbours. The centre of the mask is placed over a_{ij}. Each distance map pixel lying underneath the mask is then assigned the distance measured at a_{ij} plus the value of its own corresponding mask location.

Figure 5.9 shows examples of the masks used to compute various distance transforms; corresponding distance maps are given in figure 5.4.c and d and figure 5.10. A mask computing Euclidean distance would comprise concentric rings of equivalent, real-valued weights whose values increase with distance from the mask centre. The masks shown in figure 5.9 represent integer approximations to this ideal.

The chessboard distance transform (figure 5.9.a) is perhaps the simplest, but suffers considerable error in its approximation to Euclidean distance (some 29% [97]). The set of so-called chamfer distances [93] have more complex masks but are more accurate. The 3-4 and 5-7-11 chamfer transforms (figure 5.9.c, d) give errors of 8.09% and 2.02% respectively when compared to Euclidean distance [97].

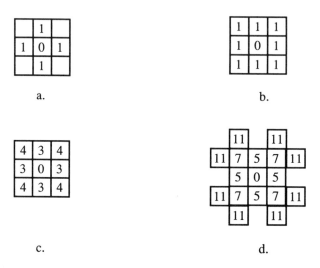

Figure 5.9. Masks used to compute various distance transforms; a) 4-neighbour, b) 8-neighbour, c) Chamfer 3-4 and d) Chamfer 5-7-11.

```
0 0 0 0 0 0 0        11 8 7 6 7 8 11       18 14 11 10 11 14 18
0 0 0 0 0 0 0        10 7 4 3 4 7 10       16 11  7  5  7 11 16
0 0 0 x 0 0 0         9 6 3 x 3 6 9        15 10  5  x  5 10 15
0 0 0 0 0 0 0        10 7 4 3 4 7 10       16 11  7  5  7 11 16
0 0 0 0 0 0 0        11 8 7 6 7 8 11       18 14 11 10 11 14 18

       a.                   b                        c.
```

Figure 5.10.a) A simple image and distance maps generated using b) chamfer 3-4 and c) chamfer 5-7-11 masks.

The masks described above may be applied in either a parallel or a sequential manner. To give an example, consider how a distance map based upon the 3-4 chamfer metric might be obtained. The first stage is initialisation in which object pixels receive 0 and background pixels receive infinite values:

$$a_{ij} = \begin{cases} 0, \text{ if } a_{ij} \in B \\ M, \text{ otherwise} \end{cases} \quad (5.5)$$

Binary Image Processing and the Raster to Vector Transform 83

where M is a very large number. If the distance transform is computed by parallel propagation of local distances, then at each iteration every pixel obtains a new value, using the following formula:

$$a_{ij}^k = \min\left(a_{i-1,j-1}^{k-1} + 4, a_{i-1,j}^{k-1} + 3, a_{i-1,j+1}^{k-1} + 4, a_{i,j-1}^{k-1} + 3, a_{i,j}^{k-1} + 3, a_{i+1,j-1}^{k-1} + 4, a_{i+1,j}^{k-1} + 3, a_{i+1,j+1}^{k-1} + 4\right) \quad (5.6)$$

where a_{ij}^k is the value of the (i,j) pixel at iteration k.

The sequential distance transform algorithm is executed during two successive image scans performed in forward (top left to bottom right) and backward fashion. The following operations are used to compute distances for object and background regions of the image A (B and C are new images).

Forward scan:

$$b_{ij} = \min(a_{i-1j-1} + 4, a_{i-1j} + 3, a_{i-1j+1} + 4, a_{ij-1} + 3, a_{ij}) \quad (5.7)$$

Backward scan:

$$c_{ij} = \min(b_{ij}, b_{ij+1} + 3, b_{i+1j-1} + 4, b_{i+1j} + 3, b_{i+1j+1} + 4) \quad (5.8)$$

Figure 5.11 shows the result of applying this transform to a simple binary image.

-11	-8	-7	-6	-6	-7	-8	-11	-12
-8	-7	-4	-3	-3	-4	-7	-8	-11
-7	-4	-3	3	3	-3	-4	-7	-8
-6	-3	3	4	4	3	-3	-4	-7
-6	-3	3	6	7	4	3	-3	-6
-6	-3	3	4	7	4	3	-3	-6
-7	-4	-3	3	4	3	-3	-4	-7
-8	-7	-4	-3	3	-3	-4	-7	-8
-11	-8	-7	-4	-3	-4	-7	-8	-11
-12	-11	-8	-7	-6	-7	-8	-11	-12

Figure 5.11. A distance map computed using the 3-4 chamfer mask for both B (positive) and W (negative) pixels.

Many variations on the distance transform theme exist. The constrained distance transform is a special case of the distance transform in which the source image consists of object, background and obstacle pixels. The values in a constrained distance map give distance along the shortest path that does not pass through any obstacle pixel(s) [95]. The distance transform for line patterns (DTLP) is a further specialisation [98,99]. The DTLP is applicable only to images in which all objects are line patterns of unit width. In the DTLP, any non-object pixel is effectively an obstacle: distance is measured not along an approximately straight line but along the line pattern. Examples of distance maps produced via the distance transform, constrained distance transform and DTLP are shown in figure 5.12.

2	2	2	2	2	3	4	4
2	1	1	1	2	3	3	3
1	1		1	2	2	2	2
1			1	2	1	1	1
1		1	1	2	1		1
1	1	1	2	2	1		
2	2	2	2	2	1	1	
3	3	3	3	2	2	1	1

a.

1	1	1	2	3		7	7
1		1	2	3		8	8
1	1	1	2	3		5	6
2				3	4	5	6
3		5	4	4	4	5	6
4		5	4				
5	5	5	4	3	2	1	
6	6	5	4	3	2	1	1

b.

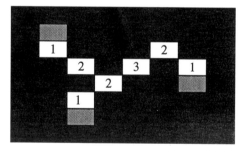

c.

Figure 5.12. Example of a) the distance transform, b) the constrained distance transform and c) the DTLP.

The DTLP assumes the input image to contain strings of data points that are one pixel wide <u>almost</u> everywhere. Broader clusters are expected only at particular types of crossing points and their neighbourhoods [98]. Data of this type is typically generated via edge detection or by thinning (see Chapter 6) binarised images of line drawings. In Toriwaki et al's DTLP [98] the ends of the data point strings are considered to be features, the remainder of the line data is non-feature pixels and the image background is treated as an obstacle. The effect of the DTLP is to assign to each pixel on the line pattern some measure of the distance to an end point. Toriwaki et al describe two forms of DTLP, which they term Type I and Type II. The latter measures the distance to the nearest end point (figure 5.13); the former, given simple figures at least, records the distance to the furthest such feature. Toriwaki et al give parallel local algorithms for the computation of both measures and illustrate their use with reference to the analysis of X-ray images of blood vessels and photomicrographs of metal components.

```
* * * * * * * *         * * * * * * * *         * * * * * * * *
* * 0 * * * * *         * * 1 * * * * *         * * 1 * * * * *
* * 0 * * * * *         * * 2 * * * * *         * * 2 * * * * *
* * 0 * * * * *         * * 3 * * * * *         * * 3 * * * * *
* * 0 * * * * *         * * 4 * * * * *         * * 4 * * * * *
* * 0 * * * * *         * * 5 * * * * *         * * 4 * * * * *
* * 0 * * * * *         * * 5 * * * * *         * * 3 * * * * *
* 0 * 0 0 0 0 *         * 2 * 4 3 2 1 *         * 2 * 4 3 2 1 *
* 0 * * * * * *         * 1 * * * * * *         * 1 * * * * * *
* * * * * * * *         * * * * * * * *         * * * * * * * *

        a.                      b.                      c.
```

Figure 5.13. a) Data and the result of applying Toriwaki et al's b) Type I and c) Type II DTLPs to that data.

More recently, the DTLP was reviewed by Ragnemalm and Ablameyko [99]. They highlight the DTLP algorithms' inability to cope with line patterns containing closed loops and/or multi-pixel intersections (features commonly found in line drawings) and propose a modified DTLP in which both terminations and intersections are considered to be features. Ragnemalm and Ablameyko further describe a sequential algorithm that computes their transform while avoiding some of the limitations of Toriwaki et al's algorithms. A key feature of Ragnemalm and Ablameyko's discussion is their questioning of the value of the whole concept of a distance map for line patterns. Their algorithm produces, as a side effect, data structures describing feature points (terminations, intersections) and the branches (strings of 2-connected pixels) which link them. This segmentation is in itself a valuable operation. Indeed, Ragnemalm and Ablameyko suggest that it may be of greater practical value than the distance maps provided by either the DTLP or MDTLP. The central tenet of [100] is that the various forms of DTLP described to date are of limited value, not because distance transforms per se cannot be usefully applied to line data, but because those proposed thus far do not reflect adequately the structure of that data.

In general, there are many paths from each element of a line pattern to a feature pixel of some given type. A major issue to be addressed during the interpretation of line data is which paths are important and which are not. It is our belief that the role of a distance transform should be to provide information that supports higher-level systems in making this decision. The distance maps generated by the Type I and

Type II DTLPs of Toriwaki et al, however, effectively ignore the graph structure of their data, choosing instead to provide information about only one path from each point; the longest or shortest to a feature point. The process intended to support decision making implicitly makes the decision. The modified DTLP of Ragnemalm and Ablameyko clearly takes more account of the structure of its data; though it too performs a segmentation, this time explicitly. It was their consideration of this segmentation which led them to question the value of their distance map.

The central feature of the Generalised DTLP (GDTLP) is the combination function

$$C_{xy}(g_{xy1}, g_{xy2}, g_{xy3}, ..., g_{xyn})$$

in which each $g_{xyi} = G(l_{xyi})$ and $l_{xyi} = L(p_{xyi})$, i.e. $g_{xyi} = G(L(p_{xyi}))$. In the above, $l_{xyi} = L(p_{xyi})$ is a measure of the length of the ith path p_{xyi} from a data point at pixel (x,y) to some appropriate feature pixel and $g_{xyi} = G(l_{xyi})$ is a function of that length. In traditional distance transforms $G(l_{xyi}) = l_{xyi}$ in general, however, $G(l_{xyi})$ may measure any useful path property. Each path p_{xyi} comprises an ordered list of pixels

$$p_{xyi} = ((x_1, y_1), (x_2, y_2), (x_3, y_3), ..., (x_m, y_m))$$

such that $x_1 = x$, $y_1 = y$, (x_m, y_m) is a feature pixel and adjacent members of the list are neighbours within the input line pattern. More formally, we define a binary-valued feature detection operator F and a binary-valued neighbourhood test N and require that

$$F(x_m, y_m) = 1 \text{ and } N(x_i, y_i, x_{i+1}, y_{i+1}) = 1 \text{ for } 1 \leq i < m$$

In general, the set of paths $P^c_{xy} = \{p_{xyi} : 1 \leq i \leq n\}$ contributing to

$$C_{xy}(G(L(p_{xyi})) : 1 \leq i \leq n)$$

is a subset of the set of all paths P_{xy} to a feature pixel from (x, y). Membership of this subset is determined by the binary-valued selection function $S_{xy}(p_{xy})$. Hence

$$P^c_{xy} = \{p_{xyi} : p_{xyi} \in P_{xy} \text{ \&\& } S_{xy}(p_{xyi}) = 1\}$$

GDTLP admits many transformations, a given transform T being defined by

$$T = \{C_{xy}, S_{xy}, G, L, N, F\}$$

Output from the GDTLP is a generalised form of distance map, which we term the Generalised Distance Map for Line Patterns, the GDMLP. In its pure form this comprises an array containing at each data pixel the value of $C_{xy}(G(L(p_{xyi})) : 1 \leq i \leq n)$ determined by T. Background pixels we assume to contain zeroes. Some transformations may compute useful information as a stepping stone to $C_{xy}(G(L(p_{xyi})))$. To enable this to be retained we further allow each pixel of the

GDMLP to be associated with a pointer to an optional secondary data structure. The utility, size and content of these secondary structures also vary with T.

We now consider one possible GDTLP algorithm, see also [100]. In what follows we assume that the input line pattern does not contain multi-pixel nodes (or clusters). Though this is not generally true, clusters are usually both small and artefacts of pre-processing; they are not important in themselves. We therefore assume that either each cluster would be replaced by a single junction node or the algorithm would be extended to copy an appropriate value to each pixel in a cluster. In either case the output would not be a true GDMLP in that not all cluster pixels would be assigned the correct value, though this is only a small deviation.

The algorithm alternates between propagating GDTLP values along curvilinear segments and combining the computed values at junction nodes. The algorithm terminates when no more changes can be made. For now we also assume:

1. $L(p_{xyi})$ is as in [98,99].

2. F marks termination points, retaining the spirit of the original DTLPs

3. $N(x_i, y_i, x_j, y_j) = 1$ if (x_i, y_i) and (x_j, y_j) are 8-connected, 0 otherwise; this is the most general form of connectivity in common use.

4. Only *direct* paths will be considered. We consider a given path $P_{xy} = ((x_1, y_1), (x_2, y_2), (x_3, y_3),....,(x_m, y_m))$ to be *direct* iff $L(p_{xi,yi}) > L(p_{xi+1,yi+1})$ for $1 \leq i \leq$ m-1. That is, the distance to the feature point at (x_m, y_m) is strictly monotonically decreasing as we move along p_{xy} from (x_1, y_1) towards (x_m, y_m). Hence we are only concerned with a particular type of smooth path, removing those which zigzag towards and away from (x_m, y_m).

The algorithm is as follows:

GDTLP Subset:
{
  Input an RxS binary image
  Apply F to image to mark features
  Initialise the (RxS) GDMLP - only pixels satisfying F are set to 1, other
        data pixels are labelled *invalid* while background pixels are set to 0

  Do {   /* A two-stage iterative algorithm */
    anychange = 0
    Do {   /* 1: propagate distance transform values along curvilinear segments */
    linechange = 0
    If (∃ an *invalid* pixel x,y which has exactly 2 neighbours AND exactly one of
                those neighbours holds a valid GDTLP value)
        {
            Compute $G(L(p_{xy}))$ from the valid GDTLP value
            Apply C_{xy} (see below) writing the result into the GDMLP at x,y
            linechange = linechange + 1
            anychange = anychange +1
        }
    If (∃ an *invalid* pixel x,y which has exactly 2 neighbours AND both of those
                neighbours hold a valid GDTLP value)
        {
            Apply C_{xy} to combine the GDTLP values associated with the
                two paths, writing the result into the GDMLP at x,y
            linechange = linechange + 1
            anychange = anychange +1
        }
    } While (linechange > 0)/* while changes continue to be made */

    Do {   /* 2: combine paths at junctions */
        nodechange = 0
        IF (∃ an *invalid* pixel x,y which has > 2 neighbours AND at least
                one of those neighbours holds a valid GDTLP value)
        {
            Apply C_{xy} to combine the GDTLP values associated with the
                valid paths, writing the result into the GDMLP at position x,y
            nodechange = nodechange + 1
            anychange = anychange +1
        }
    } While (nodechange > 0)   /* while changes continue to be made */

  } While (anychange >0)

  Display/Output the completed GDMLP
}

Figure 5.14.a shows an artificial binary image; object pixels are labelled 0, background *. Figure 5.14.b shows, as a particular instance of the GDMLP, a DMLP in which each pixel records the average length of paths from that pixel to a termination (G(lxyi) = lxyi, Cxy() = MEAN()). The DMLPs of figures 5.14.c and d are based upon a G that estimates cumulative curvature. Figure 5.14.c gives the minimum ($C_{xy}() = MIN()$) and figure 5.14.d the average ($C_{xy}() = MEAN()$) cumulative curvature of paths to a termination. The curvature measure used here is simple and based upon chain codes. Throughout the figure, S_{xy} = {set of direct paths such that the interiority of each pixel in those paths is <= the interiority of x,y} where the *interiority* of a pixel is the minimum number of line segments which must be traversed in order to reach an end point.

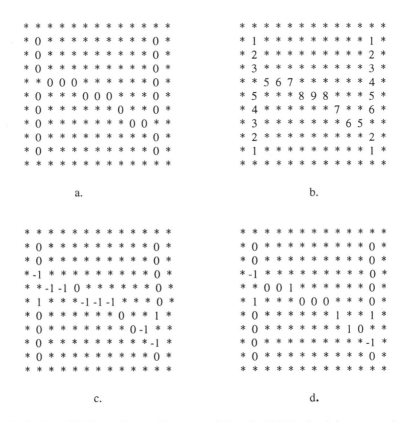

Figure 5.14. a) Artificial test image, b) mean path length DMLP, c) minimum cumulative curvature DMLP, d) mean cumulative curvature DMLP.

The GDTLP extends the DTLPs and MDTLP in two ways. First, it considers properties other than simple length. In this it owes much to the algorithm of Ablameyko, Arcelli and Sanniti di Baja [101]. Second, it recognises that there usually exist many paths from any given pixel to a feature and allows measurements of the properties of those paths to be combined in a flexible manner. It is not yet clear which of these generalisations is the most valuable. While much work remains, we believe the GDTLP will prove valuable in the segmentation and interpretation of line patterns. We see GDTLPs providing contextual information to knowledge-based, application-specific systems such as ANON (Chapter 11). Some such systems incorporate low-level processes loosely related to the GDTLP, though these typically lack any theoretical underpinning.

5.4 Mathematical Morphology

Another, parallel branch of image processing that may applied to the same tasks with approximately the same results is mathematical morphology. As a discipline, mathematical morphology was founded and developed by French scientists G. Matheron and J. Serra [102,103]. The basic idea is to probe an image with a structuring element and to quantify the manner in which the structuring element fits within the image. There are two basic morphological operations: dilation and erosion, defined in terms of set operations known as Minkowski addition and subtraction.

Let $X, Y \in A$. Minkowski addition and subtraction are, respectively

$$X \oplus Y = \{ x + y : x \in X, y \in Y \} \tag{5.9}$$

and

$$X \oslash Y = \{ h : y + h \in X \text{ for all } y \in Y \}$$

Now choose a second image Z and call it the structuring element. Then, dilation $g_Z(X)$ and erosion $e_Z(X)$ are defined as the operators:

$$g_Z(X) = X \oplus Z$$

and $\hspace{10cm} (5.10)$

$$e_Z(X) = X \oslash Z$$

respectively. Examples of these operations are shown in figure 5.15. Detailed mathematical properties of dilation and erosion are given in [104].

Two further operators are based on dilation and erosion. The operator

$$l_Z(X) = X \circ Z := (X \oslash Z) \oplus Z \tag{5.11}$$

is called the opening of X by Z. The operator

$$b_Z(X) = X \bullet Z := (X \oplus Z) \oslash Z \tag{5.12}$$

is called the closing of X by Z. Opening and closing are idempotent operators [105], which means that once they have been applied to the image, repeated applications have no further effect. Examples of the application of these operators are shown in figure 5.15.e and f.

Binary Image Processing and the Raster to Vector Transform 91

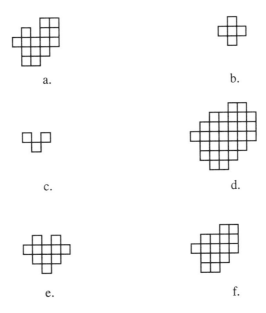

Figure 5.15. a) An image object, b) the structuring element, c) the object after erosion, d) the object after dilation, e) the object after opening and f) the object after closing.

Based on these few operators, numerous morphological algorithms have been developed. They are widely used for image skeletonisation, noise reduction, shape representation and decomposition, image segmentation, boundary detection and a variety of other image processing tasks [106-110].

5.5 Reducing Noise in Binary Images

Regardless of application, noise should as far as possible be reduced before interpretation begins. Even when subsequent operations can suppress the effects of (binary) image noise this will often require extra processing, increasing the computational resources required.

Noise can appear in a binary image in several forms (figure 5.16), these are:

- isolated pixels of the opposite value in object or background regions, often known as salt-and-pepper noise (figure 5.16.a);

- small holes in objects and/or small spots on background areas (figure 5.16.b);

- contour protrusions and intrusions (figure 5.16.c);

- distinct lines merging and single lines splitting (figure 5.16.a and b).

92   Machine Interpretation of Line Drawing Images

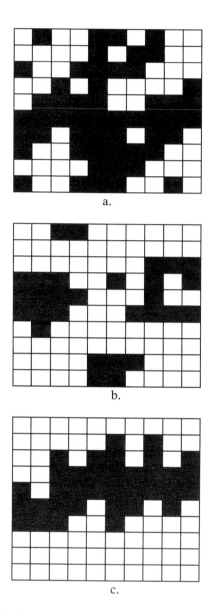

Figure 5.16. Types of noise found in binary images; see text for details. Noise has been added to a 3-pixel wide line crossing the image at an angle of 20° to the horizontal.

We now consider some generally applicable noise reduction algorithms. Perhaps the simplest are based on logical masks or filters. Each pixel a_{ij} in the binary image is examined in turn and the number and distribution of black and white pixels in a simple neighbourhood centred on a_{ij} is determined. The standard 8-neighbourhood is commonly used. Any logical test may be applied to determine the new value of a_{ij}, but in the simplest case this depends only upon the number of black and white pixels in the neighbourhood. For example, let C be a new, reduced noise image. Then,

$$c_{ij} = \begin{cases} 1, & \text{if } A_8(a_{ij}) > h \\ 0, & \text{otherwise} \end{cases}$$

where h is some user-defined threshold value. This operator can be applied independently to both object and background pixels, when two distinct thresholds would probably be used. An example of the operation of this type of noise reduction is shown in figure 5.17. It is quite common nowadays for noise removal via logical masks to be performed within the document scanner.

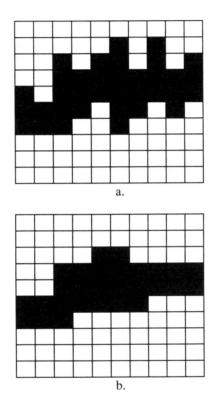

Figure 5.17. Noise removal by a 3 x 3 logical mask; a) the original and b) the cleaned image.

Further noise filtering algorithms are based on morphological operations and the distance transform. They rely on sequential shrinking and/or expanding of image objects, achieved by morphological closing and opening or via the application of thresholds to distance maps. Opening, for example, smooths object borders, breaking off narrow isthmuses and eliminating small (black) noise regions. Closing also smooths borders, bridging narrow gaps and filling in small (white) holes [105].

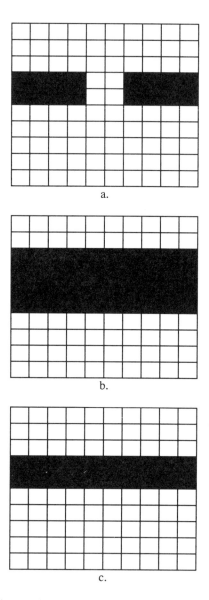

Figure 5.18. Gap closure by morphological opening; a) the original image, b) after dilation and c) after subsequent erosion.

Figure 5.18 shows the opening of a simple binary image by the 3 x 3 structuring element Z shown in figure 5.15.b. Further discussion of the use of morphological operators for noise reduction can be found in [106-110]. As different morphological operations reduce different types of noise, determination of the appropriate noise reduction strategy must be based on an examination of the noise properties of the given image.

Binary Image Processing and the Raster to Vector Transform

Distance transforms can also be used [111] to shrink and expand regions and so to reduce most types of noise. The simultaneous use of the distance transform for B and W allows one to smooth both object protrusions and intrusions at the same time. Small holes in B and W can also be filled. The algorithm proceeds as follows:

1. Compute a distance map for both object (positive values) and background (negative values) regions.

2. Apply (separate) thresholds to the object and background regions of the distance map: all pixels having distance values less than the appropriate thresholds are set to zero. This will cause a channel of zeroes to appear between the positive and negative regions of the distance map.

3. A reverse distance transform is then performed, propagating distance information away from the positive distance values (object pixels) into the channel of zeroes. Suppose a threshold t has been applied to the object regions of the distance map; contour pixels will all contain t, with recorded distances increasing further into the object region. The reverse distance transform spreads distances into the zero channel, a layer of (t-1) is added, then (t-2), (t-3), etc. down to 1. The boundary between 0 and 1 in this final distance map marks the border of the object.

The above algorithm basically shrinks the object by thresholding its distance map, then expands it again via the reverse distance transform. Background regions are shrunk simply to make space and allow the object region to expand freely. Examples of noise reduction using this algorithm are shown in figure 5.19. A threshold of 2 (appropriately signed) was applied to both maps in stage 2.

Figure 5.19. Noise removal via the distance transform; a) the original image object and b) the same object after noise reduction. See text for details.

5.6 Reducing Noise in Binary Images of Line Drawings

The noise reduction methods discussed above are applicable to any binary image. As noted in Chapter 1, images of line drawings have a number of particular properties; some arise from their content, others are related to their size.

From a computational point of view, logical mask algorithms are very fast and, perhaps more importantly, access each (central) pixel only once. Unfortunately, they are limited in the types of noise they can handle. Distance transform and morphological algorithms, on the other hand, can be used to reduce all types of noise. They, however, require the image to be scanned several times and the whole binary image and distance map to be simultaneously available. This is costly in both time and storage. A noise reduction method suited to the large scale images arising from line drawings would ideally combine these properties. It should reduce all types of noise, not require simultaneous access to the whole image, and not require many image inspections. A technique has been developed [45], which uses only one image inspection and stores a limited number of rasters in memory at any given time. The method comprises three algorithms [13], which may be used in isolation or in sequence:

1. Intelligent application of the morphological opening and closing operations. Instead of applying opening or closing uniformly to every pixel of the binary image, as is more commonly done, one or the other operator is applied to each pixel depending upon its local neighbourhood. Indeed, the neighbourhoods of some pixels may indicate that neither operator is required, in which case none is applied. The process is performed in (pseudo-)parallel during one image scan using a set of logical masks to determine the required operation and the 3 x 3 structuring element shown in figure 5.15.b. On completion of the process the image cannot strictly be said to have been either opened or closed, due to the context-dependent nature of the processing applied. The method is, however, computationally efficient and effective in the removal of salt-and-pepper noise (figure 5.20).

   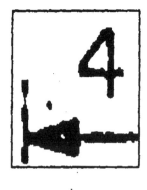

a.   b.

Figure 5.20. a) An image fragment and b) the result of salt-and-pepper noise removal from a).

2. Small spot and hole removal based on contour following. Here it is assumed that the dimensions of the minimal surrounding rectangle of significant connected components (d and h in figure 5.21) is known a priori. The height of the image stripe is then set to the minimal significant height plus one. As the stripe is swept over the image a simple contour tracing algorithm (Chapter 6) identifies black/white boundaries and measures the height and width of each spot/hole. If these values fall below the preset threshold the traced component is deleted from the image. An example of the effect of this operation is shown in figure 5.22.

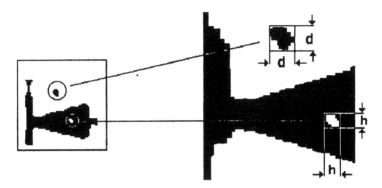

Figure 5.21. Hole and spot parameters.

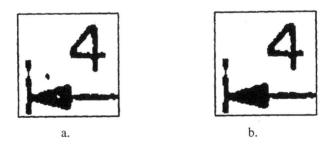

a.   b.

Figure 5.22. The removal of small spots and holes; a) a raw binary image and b) the corresponding processed binary image.

3. Contour smoothing via logical filters. A set of oriented logical filters is applied which add and/or delete contour pixels to produce a smoother object boundary. Four sets of filters, oriented at 0, 45, 90, and 135 degrees to the image co-ordinate exist, the first step in the algorithm being to determine which set should be applied to the given pixel. The length and width of the filters similarly varies within each oriented set; the length and width of the masks applied depend upon a priori knowledge, provided by the user, of the expected dimensions of contour defects. An example of the application of this method to a segment of engineering drawing is shown in figure 5.23. For some document types, specific filters can be designed which take advantage of the known characteristics of the input image. The kFill algorithm [72], for example, has been designed to reduce salt-and-pepper noise in text images while maintaining readability. A k x k window is applied to the image in raster scan order. This window comprises a central (k-2) x (k-2) core region and a set of 4 * (k-1) perimeter pixels known as the neighbourhood. All core pixels are then set to either black or white depending upon the pixel values in the neighbourhood. The algorithm is applied iteratively, one pass turning pixels black, the next turning them white.

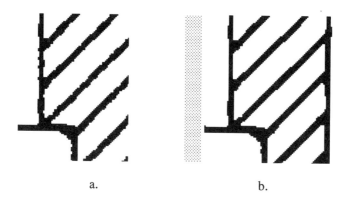

a.   b.

Figure 5.23. Contour smoothing of an engineering drawing; a) the input image and b) the smoothed version.

Three neighbourhood properties affect the decision to turn a given core area black (white), these are:

- N: the number of black (white) pixels in the neighbourhood

- C: the number of black (white) connected components in the neighbourhood

- R: the number of black (white) corner pixels

Pixel values are changed when

(c=1) AND [n>3k −4 OR (n=3k-4 AND r = 2)]

The requirement that $c=1$ ensures that connectivity remains unchanged. The other terms are designed to remove noise without rounding corners or shortening line lengths. The kFill algorithm is computationally quite expensive and, being designed for images of printed text, not always applicable to images of line drawings. Where appropriate, however, it is capable of removing significant amounts of noise without destroying the small text features upon which interpretation so often relies.

Chapter 6
Analysis of Connected Components

6.1 Nomenclature

After binarisation of the grey level image and before the construction of a vector description, object (black) regions of the binary image must somehow be reduced to unit-width pixel strings. As illustrated in Chapter 2, vector representations may describe either the centre lines of object regions or their boundaries. The processes that extract pixel strings from connected components may similarly describe centre lines or boundaries. Each representation may be achieved in a variety of ways.

In what follows we will use "skeleton" to refer to representations of the centre line of an image region. Descriptions of the object boundary will be termed "contours". The generic term "thin data" will be used to refer to both without distinction. Broadly speaking, skeletons may be achieved either by "thinning", an iterative process of removing black pixels until only a unit-width string remains, or via a "medial axis transform". Thinning produces a "thinned" binary image in which object pixels lie upon the skeleton of the original object. The medial axis transform produces a "medial axis", the locus of the centres of a set of maximal discs laid over the object. This is equivalent to the locus of local maxima of some distance measure. In many areas of the literature thinning and medial axis transform algorithms are not clearly distinguished, nor are thinned images and medial axes. The separation of skeletonisation and contouring is usually apparent.

This chapter considers the techniques outlined above. Most of the methods discussed are applicable to all types of binary image. Although the analysis of connected components has been studied for some 30 years, the number of research papers being written and algorithms being developed does not diminish. These techniques have wide application and the development of efficient, effective methods is not easy. From this point of view the key feature of line drawing images is often their large size; those working in connected components analysis for line drawing interpretation frequently find themselves almost as concerned with issues of speed and storage efficiency as they are with the quality of their output.

6.2 Contouring

6.2.1 Goals

Contouring may be applied to any image region that might usefully be represented in terms of its boundary. Contouring is a much simpler operation than thinning or the medial axis transform, especially for binary images. This is primarily because contour pixels are well defined compared to skeleton points (see Chapter 5 and below). As a result, contouring algorithms are subject to fewer requirements than thinning and/or medial axis techniques. The major requirements placed upon a contouring method are as follows:

1. the output contour should be 8- (or 4-) connected;

2. all the contours in the image should be extracted, including those internal to object regions;

3. ideally, contouring should produce a vector representation.

These requirements are quite clear, only the last warrants further comment. Many contouring algorithms simply mark contour pixels (often with chain codes) and so produce image-based representations. This is useful in, for example, the detection of symbols such as arrowheads. In line drawing interpretation, however, contouring is usually an intermediate stage in the raster to vector transformation. As the next stage in this process is typically to vectorise the contour, it would be preferable to employ an algorithm that simultaneously contours and vectorises. Many, however, produce simple strings or networks of pixels that are later vectorised by distinct processes. In what follows we shall consider contouring and vectorisation to be distinct operations.

6.2.2 Classification of Contouring Techniques

There are two common approaches to the extraction of contours from binary images: contouring in line following mode and contouring in scan line mode.

Line following or tracking is ubiquitous in image analysis. It may be applied to any form of line-like, image-based data to perform a variety of tasks (see, for example, the direct vectorisation method described in Chapter 7). Despite their diversity, line following systems are all variations on a common theme [112]; their operation can be summarised thus:

1. Scan the image to identify a starting point for tracking. This is usually achieved via a top left to bottom right raster scan of the input image. Consider the starting point to be the current pixel and mark it as having been visited; this is to prevent the tracker going round and round in endless loops.

2. Examine some neighbourhood of the current pixel, searching for new (i.e. unmarked) items that are candidates for inclusion in the developing pixel track.

Analysis of Connected Components 103

3. Select one such candidate, consider it to be the (new) current pixel, mark it and go to stage 2.

This process continues until no acceptable unmarked pixels are present in the neighbourhood of the current pixel, at which point a single pixel track has been extracted. A number of options now become available. Some systems merely resume the raster scan, looking for further, independent pixel tracks. Others extend step 2 to include storage of candidate pixels that were not chosen for inclusion in the initial track. When the end of one track is reached they then return to those stored alternatives and follow each to its conclusion. The order in which stored pixels are considered determines the nature of the tracker's search through the data. Having done this it is again usual to resume the raster scan to ensure that all candidate pixels are considered. Depending upon the nature of the input data and the criteria used to distinguish between members of the various sets of candidate paths, a wide variety of line following systems may be generated from this simple template. Most are described using the analogy of a beetle or worm crawling around the data.

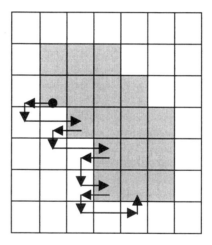

Figure 6.1. Contouring by line following.

Figure 6.1 illustrates the application of line following to the contouring problem. Having identified an initial contour point on the boundary of some object, the tracker seeks other contour pixels in the 8-neighbourhood of the current pixel, selects one and moves to it. The criteria guiding this process are usually designed to ensure that the tracker moves around the boundary in . fixed direction; either clockwise (keeping interior pixels to its right) or anti-clockwise (keeping interior pixels to its left). This method may be applied either to boundary pixels or to the horizontal and vertical line segments separating pairs of pixels. In this latter case the details of the line follower's control criteria will obviously change, although the principle remains the same.

Contour tracking is a comparatively simple technique, but often requires the entire image to be stored in computer memory. Care must also be taken when tracking around holes; control rules designed, for example, to track the boundary of a black region in a clockwise direction will not behave as expected when examining a white hole.

The scan line approach to contouring is based upon the dynamic partitioning of images described in Chapter 2 (figure 2.4). A wide, shallow window is swept down the image, contours are extracted from the image region overlaid by this window and combined with those produced at previous window locations to form a complete contour set. Ideally only one such pass over the image should be made. In effect, this type of algorithm also tracks contours. However, instead of choosing one contour at a time and seeing it to its end before moving to the next, scan line methods track multiple contours in parallel. Methods are therefore required to detect when new contours appear within the window, to extend their representations appropriately as the pass continues and to end tracking cleanly when they terminate or merge. The advantages of the scan line approach are that it requires a fraction of the computer memory of most line following systems and, particularly if only one pass is required, is very fast in operation. Although line following is now a classic image analysis technique, most current contouring systems (at least those used in line drawing interpretation) are based on the scan line model.

6.2.3 A Contouring Algorithm

To provide a detailed example of a contouring system well suited to line drawing interpretation, consider the method proposed in [45]. This is a scan line contouring system which operates upon a run-length encoded binary image to compute a vector description of object contours and the relationships between them. Below, however, we discuss only the contour extraction phase of the system.

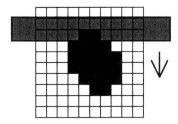

Figure 6.2. The image stripe and its movement during contour extraction.

The algorithm [45] examines two lines of the input image at a time (figure 6.2). On the basis of that examination it determines which of a number of possible situations is present in the stripe and selects a method of resolving each such situation. This resolution includes appending newly detected internal contours (those visible within the two-line window) to the appropriate external contours (which have been previously detected and are in the process of being extended). Each developing contour is stored as a unidirectional list of pixel locations, the start address of which is preserved in the description of the external contour. The five situations that might be encountered by two scan lines are:

Analysis of Connected Components

- new object;
- object continuation;
- object splits;
- objects merge;
- object ends.

As the image is analysed from top left to bottom right, these situations may be characterised as follows:

a) a new object is detected when the black run of the bottom line is fully covered by the white run of the top line (figure 6.3.a);

b) an object continuation is present when the black runs of the two lines at least partially overlap (figure 6.3.b);

c) an object split is detected when two distinct but adjacent black runs on the lower line are covered by a single black run in the upper line (figure 6.3.c);

d) conversely, when a black run on the lower line covers two separate but neighbouring black runs on the upper line objects are merging (figure 6.3.d);

e) finally, an object end is signalled when a white run in the lower line fully covers a black run in the upper line (figure 6.3.e).

Associated with each developing contour is a buffer recording descriptions of both the individual contour and its relation to others. This buffer assigns a type to each contour segment; left or right depending upon which side of the contour the image is black and on which white. It records the position of the contour within the image, its length and the dimensions of its bounding box. Also stored is the contour's internal number, this records the number of boundaries which must be crossed when moving from that contour to a background region. The external boundary of an object has internal number 0, the boundary of a hole within that object 1, the boundary of a black island within that hole 2, etc.

Buffers are merged or split, according to situation, as contouring proceeds. When a new object is encountered, two buffers (one each for the left and right branches) are selected from a free buffer stack. Each pair of branches is assigned its own identifying number that increases with the appearance of new branches. When an object continuation is detected the co-ordinates of the new contour points are added to the appropriate buffers, whose addresses are extracted from the upper line description, and the geometrical parameters are refined. Simultaneously, in the full algorithm, each branch is approximated by a polygon (Chapter 7). If an object split is detected, new buffers are created in a manner similar to that employed when a new object appears.

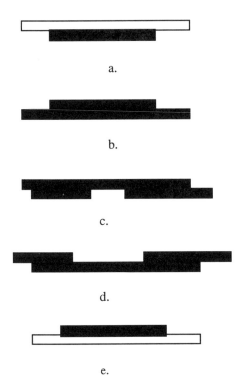

Figure 6.3. The five contouring situations.

Objects appear to merge when either tracking of an internal boundary is complete or when two branches of one contour join together. In the first instance, the information in the two relevant buffers is concatenated and passed to the output data structure. In the second, the buffer attached to the branch with the lower identifying number absorbs the data associated with the other branch. A similar process is applied when an object appears to end; this situation arises when the tracing of the external contour is complete or, again, when two branches of one contour join. Figure 6.4 shows an example of contour extraction from a segment of an engineering drawing.

Analysis of Connected Components

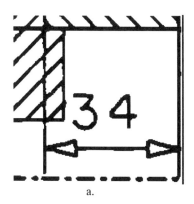
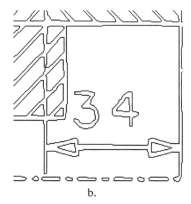

Figure 6.4. Contour extraction from a segment of an engineering drawing; a) the binary image and b) its contour.

6.3 Skeletonisation

6.3.1 Motivation

Following the conventions adopted in Chapter 5, let B be the set of pixels arising from image objects (i.e. black pixels), W the set of pixels corresponding to image background (i.e. white pixels) and S be the set of pixels comprising the skeleton of B. The following requirements are generally placed on S [113]:

1. S should have the same number of 8-connected components as B, and each component of S should have the same number of 4-connected holes as the corresponding component of B;

2. S should be centred within B;

3. S should be 8-connected;

4. each pixel in S should be labelled with an accurate estimate of its distance from W;

5. S should include all the maximal centres of B;

6. it should be possible to reconstruct B from S.

The first requirement guarantees that the topology of the initial object will be preserved. 8-connectivity is normally used for B and 4-connectivity for W (Chapter 5). Each skeleton should therefore have the same number of 8-connected components as B. Similarly, each component of S should the same number of 4-connected holes as the corresponding component of B. The second requirement stipulates that the skeleton should lie as close as possible to the centre line of the object. In a digital image it is not possible to extract true centre lines; they can only be approximated. If a given object is an even number of pixels wide, for example, its skeleton must lie closer to one border than the other (figure 6.5).

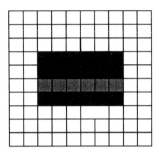

Figure 6.5. The skeleton (shown grey) of an object with even width.

While the need to retain the topology of the original object could be stated in terms of any form of connectivity, it is advantageous (as noted by requirement 3) if S is 8-connected. This implies that S will comprise a minimal number of connected pixels; were 4- or mixed connectivity to be employed, subsequent processes would almost certainly have to deal with more data.

Information regarding the width of the original object is important in line drawing interpretation; line type can often be determined by consideration of width alone. While our fourth requirement above calls for each pixel in S to be associated with an accurate estimate of local line width, this level of detail may not be necessary. It is often enough to record an inaccurate, but easily computable, approximation to line width in each pixel. In thinning, a common approach is to record the number of iterations required to generate the skeleton. Alternatively, one might retain a simple global measure; even the average width of a skeleton branch can be useful.

While, in general use, they force skeletons to be in some sense complete, for line drawing applications the last two requirements are also not strictly necessary. Information about object width is often vital, but line drawing interpretation systems do not usually need to reconstruct the initial object from its skeleton. Similarly, as long as there are enough to form a sufficiently expressive representation, it is not necessary to include the centres of all maximal discs.

Some workers active in this area would add other requirements to those given above. It might be specified (a) that line end locations should remain (approximately) fixed and/or (b) that extraneous spurs (short branches) should be minimised [72]. We would agree that it is important to minimise line shortening, though in thinning this will always happen to some extent. Similarly, there will always be some short branches, not all of which should be deleted. We prefer, however, to leave the consideration of this type of feature to subsequent processes.

Analysis of Connected Components

6.3.2 Approaches To Thinning

The classical approach to skeletonisation is thinning. Thinning is essentially an iterative shrinking process: on each iteration, each contour pixel is analysed, and, if certain removal criteria are satisfied, that pixel is deleted (i.e. turned white). Whether or not a pixel survives will depend upon the configuration of pixels in its local neighbourhood. Pixel removal continues until no further pixels satisfying the removal criteria remain, at which stage the object should have been reduced to a unit width string. The precise form of this string obviously varies with the removal criteria employed.

Pixel removal criteria, like distance transform weights, are usually expressed as two-dimensional arrays or masks. Here, however, the mask should be thought of as a template partially specifying a neighbourhood from which the central pixel should be removed. Figure 6.6, for example, shows the templates used by Arcelli et al [114]. The presence of 1s or 0s in the mask means that a given pixel should be removed if 1s and 0s appear at corresponding locations in its neighbourhood. The asterisk (*) is a wildcard; the contents of neighbourhood locations marked in this way do not affect the decision-making process. Thinning proceeds via the repeated application of masks to contour pixels. The process is usually controlled by a scan line or line following mechanism. One application of each of the chosen masks to each remaining contour pixel constitutes one iteration of the thinning algorithm. As in many thinning techniques, Arcelli et al's masks were determined empirically [90], but have been used successfully to skeletonise binary images.

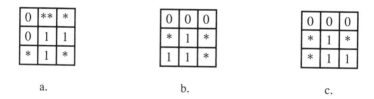

Figure 6.6. Arcelli et al's [114] thinning masks; see text for details.

Analysis and removal of contour pixels can be achieved sequentially or in parallel. The key difference here is that in parallel algorithms, pixels survive or are deleted on the basis only of the results of the previous iteration. In principle at least, all contour pixels are examined independently and in parallel and, if appropriate, modified independently and in parallel. In sequential algorithms, contour pixels are examined in some predetermined order. Deletion or survival of a given pixel is again a function of its neighbourhood, but some members of that neighbourhood will already have been considered and perhaps modified during the current iteration, while others will not.

Perhaps the best known sequential thinning algorithm is due to Hilditch [115]. Here the image is scanned from top left to bottom right and four pixel removal conditions are tested:

1. at least one black neighbour of a_{ij} must be unmarked;
2. $C_n(a_{ij})=1$ at the beginning of the iteration;
3. if x_3 is marked, setting $x_3=0$ does not change $C_n(a_{ij})$;
4. as 3, but with x_5 replacing x_3.

The first condition was designed to prevent excessive erosion of small "circular" objects, the second to maintain connectivity. The remaining two preserve two-pixel wide lines, which may easily be removed completely during thinning. Recall that $C_n(a_{ij})$ is the crossing number of pixel a_{ij} (Chapter 5).

Parallel algorithms have the advantage that they are obviously more suited to implementation on parallel machines. These methods can, however, have difficulty preserving connectivity, particularly if small (3 x 3) masks are employed. Segments two pixels wide can disappear completely within a single iteration. The standard approach to this problem is to divide each iteration into a number of sub-iterations. Each sub-iteration considers only those contour pixels that satisfy some condition. In this way one half of a 2-pixel line, for example, may be deleted in one sub-iteration, leaving only contour pixels which survive because they no longer satisfy the criteria for either examination or removal during future sub-iterations. A detailed review and classification (figure 6.7) of parallel and sequential thinning techniques is provided by Lam et al [90], who identify 1-, 2- and 4-sub-iteration parallel algorithms.

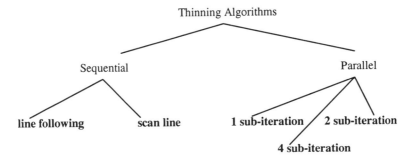

Figure 6.7. Lam et al's [90] classification of thinning algorithms.

A now classic parallel algorithm was proposed in [116]. In this method pixel p is deleted if and only if:

1. $b(p) \geq 2$
2. $X_R(p) = 2$
3. $x_1 x_3 x_5 = 0$ or $X_R(x_3) \neq 2$
4. $x_7 x_1 x_3 = 0$ or $X_R(x_1) \neq 2$.

The algorithm is also described in [117], with the added condition that $b(p) \geq 6$. This is to ensure that p has a white 4-neighbour, so that deletion of p would not create a hole. This is a one-subiteration algorithm that uses information from a 4 x 4 window. It does produce connected skeletons that are relatively insensitive to

Analysis of Connected Components 111

contour noise, but can result in excessive erosion. Recent overviews of the thinning literature can be found in [90,118-121].

6.3.3 A Thinning Algorithm

To provide an example of a thinning algorithm well suited to line drawing interpretation, consider the method developed in [122]. This algorithm displays all the major features of thinning algorithms developed for images of linework. It takes a scan line approach, scanning the image only once. It operates on run-length encoded images and it relies on 3 x 3 masks to mark pixels for removal. Finally, it uses a look-up-table for speed.

Thinning algorithms based on run-length encoded images have been proposed in [47,48]. These are fast, but can produce slightly distorted skeletons. The algorithm described here is based on the Hilditch algorithm [115], but modified for run-length encoded images. The method produces good quality 8-connected skeletons labelled with local line width.

A scan line mechanism drives the algorithm. The stripe is w+3 lines high, where w is the expected maximum width of the objects to be thinned. Lines within the stripe are numbered, bottom to top. Thinning proceeds from the first (bottom) to the last (top) line. Pixels within the scan-line are processed a varying number of times, depending on their vertical position within the window. In one application of the scan line window to the image, pixel runs in the top line are thinned completely while those in the bottom line remain unchanged. Central lines are partially thinned (figure 6.8).

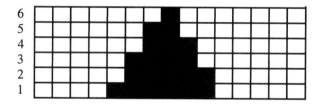

Figure 6.8. Thinning an object within the stripe.

Each line in the stripe has two descriptions: a primary, run-length description and an auxiliary, pixel array representation. The first allows easy separation of black segments requiring processing from white and completely thinned runs. This description is reformed after each iteration. The second representation (one byte per pixel) is used to record thickness and form representations of pixel neighbourhoods that can be matched to the library of 3 x 3 masks. Pixels are labelled according to a simple convention: positive values denote object thickness in the corresponding pixel; negative values denote the number of the thinning iteration on which a pixel was deleted.

Once an unthinned pixel has been identified in the run-length representation, its 8-neighbourhood is considered. A 9-bit mask is formed in which each bit corresponds to one element of the pixel neighbourhood. White pixels are coded 0, black pixels and pixels already marked for deletion during the current (kth) iteration (which are marked -k in the pixel description) are coded 1 (figure 6.9). For efficiency, pixels are considered in raster order, so the 9-bit code for the next pixel may be obtained by simply shifting the codes for the two leftmost columns of the neighbourhood to the right and introducing new codes to the left hand side. Once the 9-bit code has been formed for a given pixel it is reduced to an 8-bit mask by the simple removal of the code for the central pixel (which should always be 1). The new 8-bit code is then used to index into a 256-element look-up table.

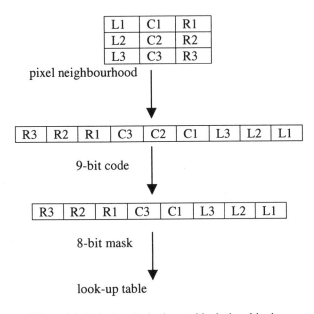

Figure 6.9. Indexing the look-up table during thinning.

The look-up table contains descriptions of all possible 3 x 3 situations associated with the actions that should be performed on the central pixel in each case. Although the process is uniform across the look-up table, two types of situation (figure 6.10) may be distinguished. In the first, all 1s in the 8-bit code represent pixels that were black when the code was constructed. In the second, 1s may represent pixels already marked for deletion. These codes may be thought of as representing parallel (figure 6.10.a) and sequential (figure 6.10.b) masks respectively (section 6.3.2). The result of the table look-up operation is to remove deletable pixels from the run-length description and update the labels stored in the pixel array. When all the pixels in a run length interval have been completely thinned, the interval is marked and not analysed further. Figure 6.11 illustrates the application of the above algorithm to an engineering drawing image. Note that every pixel in the output skeleton is labelled with a local width estimate that is not shown in this figure.

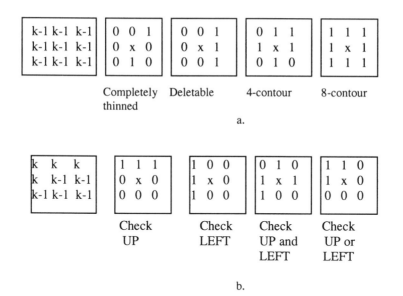

Figure 6.10. Thinning situations; a) parallel and b) sequential masks. See text for details

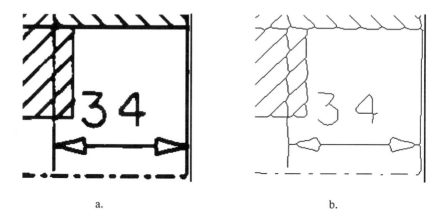

Figure 6.11. A thinned engineering drawing; a) its original form and b) the thinned image

6.3.4 The Medial Axis Transform

The Medial Axis Transform provides an alternative approach to the extraction of skeletons from binary images. Medial axis transforms rely upon distance maps obtained via the application of a distance transform. The advantages of this approach are that it supports computation and storage of quite accurate local estimates of line width and allows the original object to be fully reconstructed from its skeleton.

The medial axis transform is based on the idea that each object can be represented by a set of maximal discs. Generally, a disc(c,r), with centre $c = (c_x, c_y)$

and radius r, is a collection of pixels $p = (p_x, p_y)$ with the property that disc(c,r) = {p / $(c_x - p_x)^2 + (c_y - p_y)^2 < r^2$}. Physically, the discs are polygons approximating the Euclidean circle to some degree. Discs obtained using city-block or chessboard distances are rectangles, while those obtained via the 3,4-chamfer distance and the 5,7,11-chamfer distance have eight and sixteen sides respectively. The maximal disc is a disc that fits over the input object and is not completely overlapped by any other disc fitting the object [38,123]. The medial axis transform is completely defined by the centres of the maximal discs and their radii. This is a compact representation that compresses the image information significantly. The set of maximal discs employed may, however, substantially overlap each other. The centres of a minimal set of maximal discs can be extracted using only a little additional computation [38]. The maximal disc centres extracted from a simple object are shown in figure 6.12.

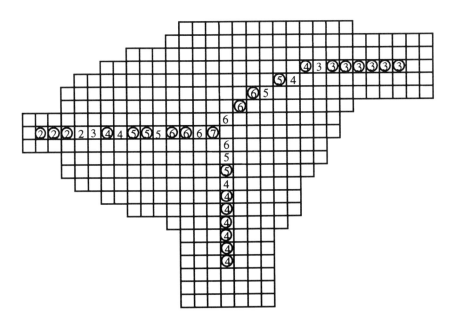

Figure 6.12. A skeleton extracted from a distance map via a medial axis transform. Note the labelling of each skeleton pixel with its local width. Ringed pixels are centres of maximal discs, other non-empty pixels are saddle/connecting points.

The medial axis is usually thin, but is not guaranteed to have a width of one pixel. Moreover, it need not be connected, even if the input object is. The medial axis transform cannot therefore be used directly for object representation and interpretation. Connected, unit-width skeletons can, however, be obtained from the medial axis. To achieve this, so-called saddle and connecting pixels must also be extracted from the distance map. Together, these three sets of pixels form a skeleton.

A saddle pixel marks a local maximum of distance in one direction on the distance map and a local minimum in another. Saddle pixels are located by counting, for each pixel in the distance map, the number of connected sets of neighbours with lower and higher distance labels. Connecting pixels link two saddle pixels and some centres of maximal discs to form connected lines. They are found by growing paths normal to the direction of the steepest gradient in the distance map, starting from an initial skeletal pixel [38,124]. The skeleton obtained may still be more than one pixel wide, so the last step in the process is to reduce it to unit thickness by employing topology and end-pixel-preserving removal operations [38]. Many papers in the literature show how skeletons can be extracted from distance maps [123-127]; an example of such a skeleton is shown in figure 6.12.

Skeletons obtained using medial axis transforms are similar to those produced via the sequential erosion of a thinning process. Medial axis transforms are usually less directionally sensitive than erosion techniques, so their output is less affected by the orientation of the object on the image plane. However, medial axis transforms are typically more sensitive to object irregularities than are iterative thinning algorithms. The resulting skeletons therefore often have more spurious branches which must be removed before interpretation can proceed [128].

6.3.5 Removing Noise from Thinned Data

After thinning or the application of a medial axis transform, it is usual to find small defects and noise in the output skeleton. This is typically the result of noise on the initial image and/or imperfections in the skeletonisation process. Pre-processing the binarised image (Chapter 5) may help reduce skeleton noise, but is unlikely to erase it completely. Noise may be reduced either immediately, by examination and modification of the skeleton, or later, after vectorisation. Removing noise from the skeleton saves the time and effort involved in vectorising spurious data and can reduce the risk of an erroneous vector description being produced. However, it typically requires the image to be accessed and inspected several times, which for large-scale drawing images can be expensive. Removing noise from the vector representation may be easier; the vectorisation process adds information that enables erroneous or spurious data to be more readily identified. The cost of this, however, is in the computational resources required to make it available. In what follows we briefly consider methods of noise removal from skeleton data; the removal of noise from vector descriptions is discussed in Chapter 7.

It is worth noting at this point that all the noise removal methods in common usage are empirically derived and somewhat ad hoc. To our knowledge, no formal model is available of the noise process at work in line drawing interpretation. As a result, no rigorously defined noise removal techniques exist. Instead, those working in this field typically gather experience of the types of noise to be expected from the application of various techniques to various image types, and design application- and technique-specific methods of dealing with them. Several types of skeleton defect, however, are commonly observed. These are (figure 6.13):

- small gaps between the end points of successive segments;

- additional short branches (node-end) emanating from skeleton nodes;

- neighbouring nodes connected by implausibly short arcs;

- spurious, free-standing short loops;

- spurious isolated segments (end-end).

Figure 6.13. a) Skeleton defects and b) their removal.

Noise is removed from the skeleton by a process known as pruning. The operation is sufficiently common that some [124,127] regard it as an integral part of skeletonisation. Pruning usually deals only with peripheral branches, i.e. those delimited by an end point. Pruning may involve either shortening or complete deletion of a peripheral branch. The elements of the branch are checked against a given pruning condition; if that criterion is satisfied the branch is pruned in some way. Several pruning criteria have been successfully employed [129]:

1. Branch length can be estimated by determining the number of pixels constituting the branch. Peripheral branches are then entirely removed if their length is below an a priori fixed threshold. In other applications, where skeletons are computed from thick regions, shortening a given branch may produce a more appropriate skeleton; in line drawing interpretation the elongation of the regions processed means that there is rarely any need to shorten branches. Most are simply deleted.

2. The area of a branch can be estimated by taking into account its length and local width measures. Area is then also compared with a threshold to determine whether the branch should stay or be deleted. Slightly more advanced techniques also compare the area of the branch with that of the whole object before making a decision.

3. Branch elongation (length/width) can be also estimated and used in a similar fashion.

Other criteria are also in use, though all reported pruning algorithms rely upon the computation and thresholding of parameters such as these. As noted above, the details of these operations vary considerably with the properties of the original image and the processes applied before and after skeletonisation. The above examples are merely intended to give the flavour of this area of drawing interpretation.

One powerful technique that might be applied to this task is the distance transform for line patterns (DTLP, Chapter 5). This supports both decomposition of the skeleton into parts and computation of those parts' parameters [101]. Based on those parameters, branches that satisfy some removal conditions (below threshold length or width, within a certain distance of a particular type of node, etc.) may be detected and deleted. The use of the DTLP in this area is a current research topic.

6.4 Grey Level Skeletonisation

The linear process sequences described in figures 2.1 and 5.1 represent by far the most common approach to line drawing interpretation and raster to vector transformation respectively. It is not, however, unusual to find systems that attempt to produce improved or more efficient results by integrating the various processes in different ways. One example is the body of work on grey level skeletonisation, which aims to move directly from a grey level image to a skeleton description without a distinct binarisation stage. As there is always some loss of information during thresholding, interest is increasing in the possibility of extracting drawing features directly from the grey level image.

Several grey-scale skeletonisation techniques have been developed [130-136]. These may be divided broadly into three groups. The first view the grey level image as a three-dimensional terrain, or digital elevation model, with each pixel's grey level providing its associated "height". Topographic features of this terrain, usually ridge lines, are sought by considering the first and second partial derivatives of the (assumed) continuous surface representing the image [136, 137]. The second group relies upon the computation of distance maps and a search for skeleton pixels within them [132]. The third and most developed set of algorithms is based on ideas from binary thinning [130,131,133,134]. They typically begin by inverting the image (black becomes white and vice-versa), then produce an 8-connected skeleton by iteratively eroding the grey level pattern, proceeding from lower grey levels towards higher ones.

Figure 6.14 shows the result of applying the grey-level thinning algorithm described in [132] to a segment of engineering drawing. This is a two-stage algorithm; the first extracts skeletons while the second deletes structural noise from the skeletons by exploiting semantic knowledge of the problem domain. The skeletons produced are generally of good quality and correspond well to ridges in the inverted grey level image.

Grey level skeletonisation by iterative pixel removal considers neither the topographic (ridges, peaks, saddle points, etc.) nor other pixels necessary to guarantee skeleton connectivity. Skeleton branches are, however, usually located either along ridges, when these exist, or centrally placed within constant grey level regions. At the beginning of each iteration the current set of contour pixels is partitioned into single grey level sets Within each iteration, pixel removal criteria are applied to the members of each set in turn, the sets being ranked in ascending order of grey level. One iteration of the algorithm is complete when all pixel sets have been examined.

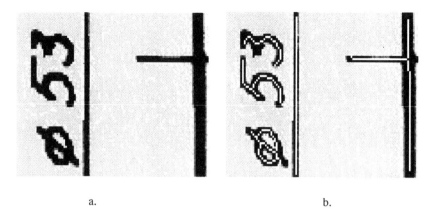

a.   b.

Figure 6.14 a) A segment of a grey level image of an engineering drawing and b) the skeleton produced by the grey level thinning algorithm of [132]

Skeletons produced in this manner suffer from defects similar to those found in binary skeletonisation. The difference here, of course, is that grey level information is available. The most frequently found forms of structural noise are:

- short branches whose grey levels are either close to the background grey level or increase with distance from the skeleton (in an inverted image one would expect branches to become darker further from the skeleton);
- branches which connect two high grey level regions through a low grey level region.

A variety of tests are applied to the skeleton to determine which branches should be pruned, which merged etc., full details are given in [138]. In essence the process is similar to that described in section 6.3.5. The availability of grey level information makes the identification of structural noise a little easier and its removal a little more reliable, but the techniques used are still derived empirically to suit the task at hand.

Chapter 7
Vectorisation

7.1 Approaches To Vectorisation

The final step in the raster to vector transformation is to move from a skeletonised or contoured binary image to the set of primarily geometrical primitives that make up the vector descriptions outlined in Chapter 2. As always, the detailed properties required of a vector description will depend upon the type of drawing being processed, the planned sequence of subsequent operations and the overall goals of the application. However, possibly because vector descriptions are closer to those used by humans in their interpretation of line drawings than the pixel-based representations considered previously, human observers are generally skilled in evaluating the quality of vector output. If a vectorisation system is to be accepted into everyday use the geometrical primitives it presents to the user must be of the right type (straight lines must be described by straight lines, circles by circles, etc.), fit the underlying skeleton or contour data accurately and be produced reasonably quickly. A fairly standard trade-off exists between these last two aims.

Implicit in the vectorisation process is a grouping and/or segmentation operation; vectorisation systems not only fit lines, curves, etc. to skeleton or contour pixels but must also decide to which groups of pixels these primitives should be matched. Again implicit in this grouping process is the recognition of pixels which are members of more than one such group (branch or node points, Chapter 5) and those which mark the free ends of line segments (end points, Chapter 5). One important choice which must be made in the design of a vectorisation system is how much of this information is to be made explicit and at what stage in the interpretation process? This will depend to some extent upon the use to which the vectors are to be put.

Underlying the thin data is a graph structure (figure 7.1) comprising branch, end and connectivity pixels. This could be made explicit at the pixel level; each pixel may be assigned a label describing its role within the graph and some reference (e.g. chain codes) to those pixels to which it is connected (figure 7.1.b). Subsequent vectorisation processes would then inherit and overlay geometrical descriptions on this structure. The final data structure might be quite complex, but would allow vectorisation and other, higher-level, interpretation processes to revisit and perhaps reconsider the underlying pixel chains as necessary.

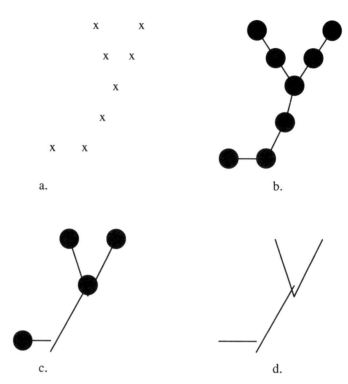

Figure 7.1. a) A simple skeleton, b) its pixel-level graph structure, c) its vector-level graph structure, d) an unstructured vector representation: nodes are only implicit here.

Alternatively, the graph structure could be made explicit only at the vector level (figure 7.1.c). Arcs in the graph would be series of vectors rather than simple pixel strings. In this approach information regarding the detailed structure of the thin data is effectively discarded and any subsequent processes that required access to individual pixels would have to search the earlier, raster-based, representations for the necessary information. If, on the other hand, such access is not required, maintenance of a pixel-level graph structure is an unnecessary computational expense.

Finally, the designer of a vectorisation system may choose not to make the underlying graph explicit in any data structure. Instead, the vectorisation process could search the skeletonised or contoured binary array for groups of pixels which form acceptable vectors and simply write each vector independently to a flat (i.e. unstructured) file (figure 7.1.d). This approach is particularly suitable if the vector data is simply to be displayed, perhaps as a backdrop to manual interpretation. The representation may also be considered appropriate for automatic interpretation systems. It might be considered that processes seeking very specific higher-level entities, dimension sets and cross hatching in engineering drawings, for example, are better placed to determine the correct graph structure for their purposes than a lower-level, more generally applicable, vectorisation algorithm.

Vectorisation

The reason for introducing these different possibilities here is that the style of vectorisation algorithm employed depends to some extent on when, and if, structural information is to be included. If a pixel-level graph structure is created during, or at the end of, the contouring or skeletonisation stage it seems reasonable to assume that the subsequent vectorisation algorithm will take advantage of the available information and consider each arc of the graph in turn. This approach we term global vectorisation, as the vectorisation algorithm has simultaneous access to each pixel in the given string. Global vectorisation methods are discussed in detail in Section 7.2, but may be divided broadly into two types. The first set of iterative methods focuses on the approximation of the input pixel string by some geometrical primitives (Section 7.2.1). Such methods only segment the string into substrings when this must be done to produce a sufficiently accurate or otherwise satisfying geometrical representation. The second set of global vectorisation techniques is feature-based (Section 7.2.2). These focus upon the identification of discontinuities in the pixel string, and only later fit geometrical primitives between these discontinuities. Finally, there exist some hybrid methods that combine these two approaches (Section 7.2.3).

If either a vector-level graph is to be created, or no structural information is to be made explicit, it makes more sense to employ one of the many local vectorisation methods in existence. These do not have simultaneous access to the entire pixel string; instead they access the thin data sequentially, constructing and updating a vector description on the basis of local segments of thin data as they go. Local vectorisation methods traverse the thin data in some fixed order, usually taking either a line following (Section 7.3.1) or raster scan (Section 7.3.2) approach. One could, of course, use a raster scan or line following technique to extract each pixel string in turn and apply a global vectorisation method without making a full pixel-level graph structure explicit. This approach would, however, be wasteful of computational resources; to reduce the time taken to produce a vector description, local vectorisation systems generally access each pixel only once. Whatever the details of an individual vectorisation system, the broad classification of vectorisation techniques introduced here provides a useful framework upon which to base our discussion of currently available algorithms. This classification is summarised in figure 7.2.

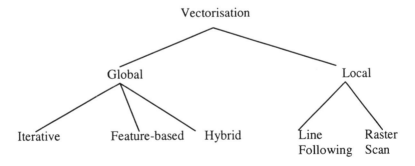

Figure 7.2. A broad classification of vectorisation methods

7.2 Global Vectorisation Methods

7.2.1 Iterative Methods

The great majority of global, iterative line approximation methods are derived from the original "iterative end-point fit algorithm" of Urs Ramer [139]. Ramer's algorithm takes a top-down approach to the geometrical description of pixel strings, proceeding as follows:

1. Hypothesise and compute the parameters of a straight line between the two end-points of the input pixel chain.

2. Compute some measure of the goodness of error of the fit between the proposed line segment and the original data.

3. If the error level is below some predetermined threshold, accept the proposed line as a sufficiently accurate representation of the input data and terminate.

4. Should the error level render the proposed line unacceptable, identify the data item that lies furthest from the line segment and break the pixel string into two at that point; apply the algorithm to each of the resulting substrings.

The iterative end-point fit algorithm systematically divides an initial line segment into smaller pieces until each element of the original pixel chain is associated with a line segment which describes the local neighbourhood to the required level of accuracy. Figure 7.3 illustrates the operation of the technique.

Many variations on this simple theme have been proposed and employed. The original algorithm, for example, simply took the line between the end-points as its initial description. This is simple and computationally cheap to implement, but the resulting line is necessarily sensitive to errors and noise in end-point position. As the end-points of pixel chains are either termination or branch/node points in a pixel-level graph structure, such effects are common. If the pixel chain was produced by thinning, for example, line intersections and terminations are the worst features upon which to base any geometrical calculations (Chapter 6). For this reason, many who base their methods on Ramer's original method prefer to use a slower, but more reliable global line fitting technique, minimising the mean squared error, for example. The results obtained after incorporating such techniques are usually higher quality and, although line fitting takes longer, often achieved in fewer recursions.

Any reasonable error measure may be used to decide whether or not a given segment is acceptable and again there is a trade-off between computational cost and reliability. Many use the mean squared error between pixel and line location, taking the perpendicular distance from a given pixel to the line as its contribution to the total error. Others simply find the maximum such distance and apply a threshold. In Ramer's original algorithm maximum perpendicular distance was also used to identify the point at which the pixel string should be segmented. Once again, any distance measure deemed appropriate for the task at hand may be used. Indeed, the selection of the single segmentation point needed by the algorithm may be achieved via any criteria – distance from the line may be only one component of the test performed or need not be involved

Figure 7.3. The operation of Ramer's iterative endpoint fit algorithm; a) the first step and b) the second step. See text for details.

Although named the *iterative* end-point fit algorithm by its inventor, the method described above is really recursive. An alternative class of algorithms based upon the so-called minimax approach is closer to the classical idea of an iterative algorithm.

Minimax algorithms seek line segment approximations that minimise the maximum (usually perpendicular) distance between the pixel chain and the approximating line segment [140]. These algorithms typically start from one end of the input chain, taking the line between the first two points as their initial geometric description. The remaining pixels in the string are then considered in turn. Each is added to the developing geometrical primitive and the parameters of that segment re-computed. This process continues until the maximum distance between the segment and the original data exceeds a threshold (figure 7.4). When this happens, the line segment available at the previous iteration is added to the developing geometric description and the process is restarted from the next pixel in the string. The parameters of each segment are therefore developed iteratively, as is the overall description; a series of geometrical primitives is produced as the focus of attention moves along the pixel chain.

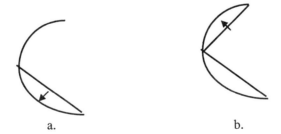

Figure 7.4. The minimax approach.

It will have been noted that minimax algorithms effectively adopt a line following approach to vectorisation, and could be considered local, rather than global techniques. This is true; some minimax methods can and have been adapted and applied to raw pixel data. We should stress, however, that minimax algorithms generally require simple, individual pixel strings to be input. A truly local algorithm would be able, indeed would be forced, to decide which pixel sets constitute significant lines. To be successful it would have to identify nodes, terminations, etc. and implicitly at least extract the major components of a pixel-level graph structure. Minimax algorithms do not attempt this, they simply use a line following strategy to iteratively describe a pre-prepared pixel chain.

All line approximation algorithms, regardless of processing strategy, are subject to a number of practical requirements. They must be sufficiently fast, immune to noise and stable over changes in the position, orientation, etc. of the pixel data. Variations in the appearance, within the image, of the input pixel string can cause approximation to begin from a different point: minimax algorithms, for example, may start from either end. The geometrical description produced must be independent of the order in which the data is considered.

Line approximation algorithms are often divided into those approximating the data by first-order polygons and those employing higher-order curves. Polygonal approximation remains the most commonly used approach within the line drawing interpretation community [141-145] although some systems use circular arcs, higher-order curves or spline functions to approximate line-drawing objects [146-150].

Systems based on higher-order curves and/or splines usually produce more geometrically accurate descriptions of the input data. They are, however, computationally quite expensive and can be difficult to implement. Practically all commercial vectorisation systems rely on polygonal approximation for object representation. Some of the approximation methods based on higher order curves are discussed in [148,149]. The application of spline approximation to the representation of cartographic and engineering drawing objects is considered in [148,151,152].

7.2.2 Feature-Based Techniques

Instead of searching for substrings of the input pixel chain which are smooth and regular enough to be described by simple geometric primitives, feature-based methods attempt to identify points at which the local properties of the data suggest the presence of a discontinuity in the underlying curve. Geometric approximation is applied, later, to the substrings linking pairs of discontinuities. Feature-based curve segmentation/description algorithms are common in the wider field of image analysis and machine vision, but less frequently used in line drawing interpretation.

Discontinuities are usually identified by consideration of local curvature; most systems store the pixel chain as a linked list of simple elements, each initially recording only the image co-ordinates of the pixels concerned. Local estimates of curvature are then computed from the local neighbourhood of each pixel in the chain and added to the list elements. The simplest way to compute curvature is to take each pixel in turn, identify the two pixels immediately connected to it in the pixel string and fit a circle through the resulting three points. Curvature is then the reciprocal of the radius of that circle. This method provides a reasonable estimate of curvature as a function of arc length (distance from one end on the pixel string), although errors associated with end- and branch-points in the pixel-level graph can make these estimates unreliable near the ends of the chain.

Once local curvature estimates are available, extrema (peaks and troughs) are generally sought in the curvature function. This is done by examining each list element in turn, checking that it is an extremal point by comparing its curvature value with those of its immediate neighbours and, if so, applying some threshold to determine whether or not the detected discontinuity is significant. The peak detection process is often called "non-maximal suppression" in the machine vision literature. Thresholds may simply be supplied by the user or determined using the types of method discussed in Chapter 4. Pixels considered to represent significant curvature extrema are marked, usually by setting a bit in the appropriate list element, and the process moves on. Examples of curvature plots and the detection of extremal points are given in the next section.

The above process identifies curvature extrema, points at which the local tangent to the input curve changes direction sharply. In many line drawing interpretation applications this is the primary, if not only, type of discontinuity to be detected. Some curve segmentation/description systems, however, also seek extrema in the first derivative of curvature (usually estimated by simple differencing of neighbouring curvature values).

Before estimating curvature and/or its derivative, many systems apply some noise reduction or smoothing operation(s) to the raw pixel co-ordinates. Local weighted averaging of pixel co-ordinates is commonly applied and, if used with care, reduces the number of false positive discontinuities detected.

7.2.3 Hybrid Approaches

To illustrate the operation of iterative and feature-based techniques and provide an example of a hybrid system combining aspects of both approaches, we now consider the algorithm of Pridmore et al [153]. Although this algorithm was developed for use in 3D machine vision, to segment and describe 3D point strings produced by an edge-based binocular stereo vision system [154], the method can equally well be applied to 2D line drawing data.

Like the iterative end-point fit algorithm, Pridmore et al's method is recursive, fitting progressively shorter straight and/or circular segments to the input string until an acceptably accurate representation is formed. On each recursion, however, extrema in curvature and its first derivative are used to identify multiple points at which the string is segmented. A feature-based component therefore provides heuristic guidance to an "iterative" contour description process. Input data is first smoothed to reduce noise by Gaussian weighted averaging of the data point co-ordinates and curvature and first derivative of curvature estimates computed at each point. The algorithm then proceeds as follows (for ease of explanation we omit details relevant only to 3D data):

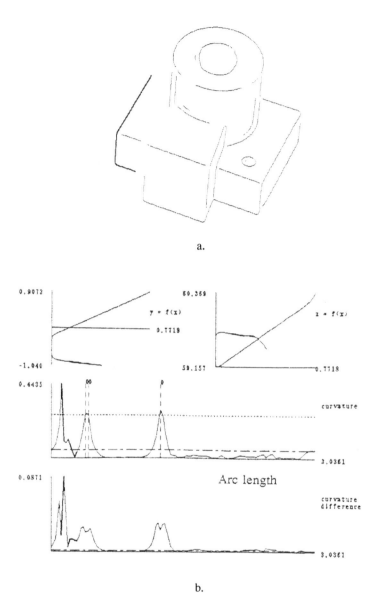

Figure 7.5. Recursive segmentation of a 3D point string; a). 3D point strings obtained from a binocular stereo vision system and b) the result of applying the algorithm of Pridmore et al to the highlighted (thick) string. The raw 3D (x,y,z) data is shown by the top two plots, which give y and z respectively as functions of x. Plots of curvature and its first derivative as functions of arc length are also shown. Horizontal dashed lines show thresholds employed, vertical dashed lines denote segmentation points. Note that the sharp peak to the far left of the curvature plot is ignored, despite its height, as it arises from horizontal data which is always unreliable in binocular stereo.

Vectorisation

- Processing begins with a call to a contour description process which fits a straight line and circular arc to the input string using a least-squares approach. Should the string be adequately described (i.e. if the residual error associated with either of these descriptions is below threshold) processing terminates. If not the string is segmented as detailed below.
- First derivative of curvature estimates are first thresholded at 90% of their maximum value, supra-threshold data being tagged as possible segmentation points. The use of 90% of the maximum means that several features are usually detected, but any that are tagged are highly likely to be true extrema.
- Tags are removed from any points that are sufficiently close to extrema in curvature that they may be considered artefacts of those curvature extrema. Details of this side-lobe test are given in [153].
- If no tags remain or all the substrings created by this operation are below a threshold length, extrema in curvature are sought. A threshold is again set at 90% of the maximum value and supra-threshold values tagged.
- If no acceptably long segments result, the algorithm terminates and no contour description is provided, otherwise long substrings are passed to the contour description process. Any that are not adequately described are further subdivided by recursive application of the segmentation procedure.

Figure 7.5 illustrates the method, which has been successfully applied to a wide variety of two- and three-dimensional line data. It should be noted that no attempt is made in this algorithm to accurately locate discontinuities; thresholding at 90% of the maximum of some curvature-related property effectively removes the region around the extrema from further consideration without the need for an explicit non-maximal suppression stage. Also, the aim of this system is to provide only reliable descriptions, rather than a best description of all the data. This means that some areas of the data will not appear in the final data set. In the domain for which the algorithm was intended, recognition of 3D objects, this is a reasonable position to adopt. It may be less appropriate in line drawing interpretation applications, though the algorithm only rarely fails to describe more than 90% of its input data.

7.3 Local Vectorisation Methods

7.3.1 Line Following Methods

As their name implies, line following methods trace around (usually thinned) binary raster data producing vector descriptions as they go. The goal of such systems is to identify strings of geometrical primitives bounded by feature pixels (Chapter 5). This can be quite a complicated task; many different permutations of closely spaced and often connected lines and feature points must be considered. To avoid this complexity, a common approach is to extract feature pixels before commencing vectorisation. This is done during an initial image scan and based on connectivity and/or crossing numbers (Chapter 5). Any feature pixels found are usually distinguished by type and stored in lists of end-points, nodes etc. During vectorisation these lists provide starting points to line approximation algorithms

[151,155,156] like the minimax methods described above. Instead of polygonal approximation of segments, circular arcs and general conics may be identified before the remaining segments are approximated by polygons. An example of this approach is given in [157].

This type of local vectorisation algorithm is comparatively easy to implement and control due to its fairly simple two-stage structure. When all feature pixels have been visited, all segments must have been traced and so the procedure terminates. Note that while the approach records the natural structure of the data it does not create a full-blown graph structure.

7.3.2 Raster Scan Techniques

While line following approximation algorithms are simple to understand and apply, most currently available local vectorisation algorithms adopt a raster scan approach [122,158]. The prime advantages of raster scan algorithms are their high speed and limited memory requirements; vectorisation is usually performed in a single image scan with a comparatively narrow window. In what follows, we describe a local raster scan vectorisation algorithm that uses a modified run-length image representation; each raster is represented by an ordered list of the x co-ordinates of black (white) pixels, which have white (black) pixels on their left (Chapter 6).

Input to the algorithm is a labelled skeleton in which every run contains a label corresponding to the thickness of the original image object. The skeleton is obtained via the thinning algorithm described in Chapter 6 and the vectorisation algorithm is similar in style. The method comprises two basic processes, situation extraction and segment following, performed simultaneously on three image lines held in a buffer.

During the first process, each pixel's local context is determined by analysing its 8-neighbourhood. Simultaneously, the crossing number C_n of the central pixel is computed. As noted in Chapter 5, a pixel is considered to be an end point iff $C_n = 1$, connective iff $C_n = 2$ and a node iff $C_n > 2$. In this last case, C_n also records the number of branches meeting at the pixel concerned. Analysis of the neighbourhood may reveal any of the following situations: line beginning, end of line, continuation, lines merging, lines branching, a node, or an isolated point (figure 7.6). Codes representing all of these situations are stored in a look-up-table which specifies the action to be taken when each is encountered.

The possible situations are divided into two groups: those in which the length of the current run is equal to one pixel (figure 7.7.a), and those in which it is more than one pixel (figure 7.7.b). In the first case, pixel type is defined immediately, the key here being that runs of one pixel are normal to the direction of the scan line, while in the second it is necessary to consider each pixel in the run in turn and determine how it should be processed.

Vectorisation

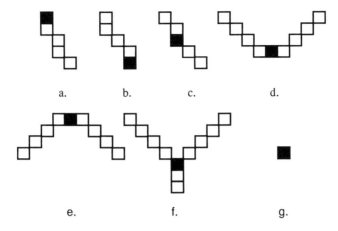

Figure 7.6. The vectorisation situations: a) line beginning, b) end of line, c) continuation, d) merging, e) branching, f) node, g) isolated point.

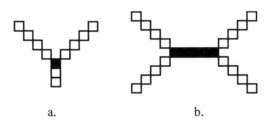

Figure 7.7. Two situations in which a) the length of the processed run is equal to one pixel, and b) it is more than one pixel.

As soon as the context of the first black pixel is determined, the second process (segment following) starts to record the details of segment relations and to prepare segment information for inclusion in a vector database. Starting from classified pixels and feature points, segments are traced during the downward motion of the stripe. Special buffers (one per segment) are initialised. These are intended for assembly, each storing information about one particular segment. During the scan, buffers may be merged or split depending upon the situation encountered. When, for example, two segments are merged together, information is transferred from one buffer to another. As a result, each image object is divided into sub-objects at its feature points and ultimately represented as a graph. As each segment is recorded in the buffer segment approximation is performed. An example of a vectorised section of an engineering drawing is shown in figure 7.8.

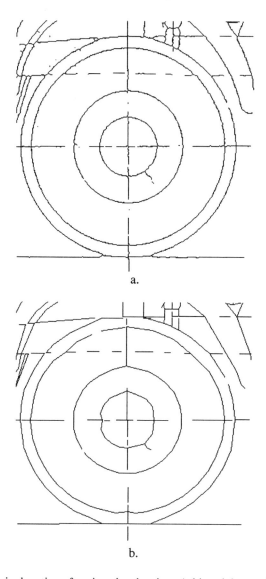

Figure 7.8. A vectorised section of engineering drawing; a) thinned data and b) vector data.

7.4 The Hough Transform

All of the vectorisation techniques discussed so far, throughout their operation, consider their input data only in image space, i.e. as collections of points and/or features lying on the image plane. A radically different approach is taken by techniques based upon the Hough transform [159]. The Hough transform is a pattern recognition technique which, given a suitable parameterisation of the construct sought, determines how many instances of that construct appear in the input data and

Vectorisation

approximates the parameters of each such instance. The Hough transform might form the basis of either a local or a global vectorisation technique, according to the meaning associated with these terms above.

The Hough transform is best explained with the aid of an example. Let us assume we wish to identify straight segments in a set of point co-ordinates provided by binarisation and thinning of a line drawing image. We therefore begin with a collection of points (x_i, y_i), on the image plane, which have a high probability of lying on the desired straight lines. We now require a parameterisation of the construct we wish the Hough to locate, i.e., an equation describing a straight line. Following the original transform [159], we shall use the standard form

$$y = mx + c \qquad (7.1)$$

in our example. Here the slope of the line m and its intersection with the y-axis c parameterise any given straight line. Given these basic components one could adopt a template matching approach, hypothesising the existence of particular lines by selecting parameter pairs (m, c) and testing those hypotheses by seeking sets of (x_i, y_i) which satisfy the constraint the hypothesised parameters impose via equation (7.1). This would, however, be a computationally expensive technique. The Hough transform approaches the same problem from a different direction. Instead of asking which of the set of possible parameter pairs (m, c) represent straight lines that actually appear in the image, it asks which lines <u>could</u> pass through each given (x_i, y_i).

Consider a particular skeleton point (x_0, y_0). Substituting into equation 7.1 gives

$$y_0 = mx_0 + c \qquad (7.2)$$

where (x_0, y_0) are known and (m, c) unknown. Rearranging equation 7.2 gives

$$c = -mx_0 + y_0 \qquad (7.3)$$

the equation of a different straight line, parameterised by (x_0, y_0) and lying not on the image but in a new space indexed by m and c. This second space is known as a parameter space, because its indices are the parameters of the objects sought by the Hough transform. Each point in this space represents one possible image line. Each point satisfying equation 7.3 represents one possible line through (x_0, y_0) on the image plane. The parameter space line described by equation 7.3 represents the set of all image lines passing through (x_0, y_0).

Suppose now that we have two skeleton points (x_0, y_0) and (x_1, y_1). Substituting and rearranging as before generates two lines

$$c = -mx_0 + y_0$$
$$c = -mx_1 + y_1$$

in m, c space. These will intersect at some point (m', c') which specifies the parameters of an image line through both (x_0, y_0) and (x_1, y_1). The situation is illustrated in figure 7.9. It follows that all data points in the image plane which are collinear with (x_0, y_0) and (x_1, y_1) will also generate lines in (m, c) space which pass through (m', c'). The Hough transform exploits this relationship, mapping all (or an appropriate subset of) the input data into parameter space and seeking points in that space through which many lines pass. The argument for this method is that the more lines passing through a given point (m_i, c_i), the more likely it is that a line with those parameters exists on the image plane.

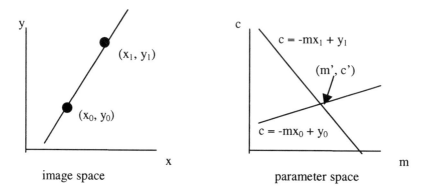

Figure 7.9. Image and parameter spaces in the Hough transform.

The Hough transform proceeds as follows:

- Choose appropriate minimum and maximum values for m and c.

- Form an accumulator array A[m, c] quantising the appropriate region of m, c space into digital "buckets".

- Initialise each accumulator to 0.

- For each image space data point (x_i, y_i), increment all members of the accumulator array lying on the appropriate line in (m, c) space; i.e. A[m, c] = A[m, c] + 1 for all m, c satisfying $y_0 = mx_0 + c$.

- Identify significant local maxima in A[m, c] and read off the corresponding estimated line parameters.

The Hough approach can in principle be employed using any parameterisation of the desired construct, though some may be better suited to the task than others. The slope-intercept format discussed here, for example, is rarely used in practice as m is infinite and c undefined for vertical lines. Hough transforms for straight lines are more commonly based on a (ρ, θ) space [160], where

$$\rho = x.\sin\theta + y.\sin\theta$$

Using this method each (x_i, y_i) generates a sinusoidal curve in parameter space, otherwise the process is unchanged. The Hough transform has also been used to detect circles [161,162], ellipses [163,164] and generalised to arbitrary contours [165].

The Hough approach is attractive because it is reasonably insensitive to both noise and occlusion of the constructs it seeks. If part of a line is obscured, for example, the remainder may reasonably be expected to produce a noticeable, if smaller, local maximum in parameter space. It should be noted, however, that the output of the Hough transform is simply a set of parameters for each object detected. In the straight line Hough, for example, only (m,c) or (ρ, θ) are estimated. Although the image data is known to lie on an estimated line, no information is provided as to where on that line the data points appear. Moreover, the data points may not be connected; a set of N collinear points constituting a single vector will produce the same response as N points on a similarly parameterised dashed, dotted or chained line. If the end-points of the constructs detected by a Hough transform are required, further processing is necessary. This may be considered either a strength or a weakness of the technique, depending on the application (see Chapter 10).

7.5 Direct Vectorisation

The principle of least commitment [33] advocates setting difficult decisions aside until sufficient information is available to allow them to be made with a reasonable chance of success. This results in interpretation systems in which each component only performs that subset of its given task that it can achieve with the knowledge and data which is naturally available at that stage in the interpretation. Thresholding algorithms built in this way, for example, only make use of grey level and other localised measures of image intensity distribution. Such systems are generally termed bottom-up. An alternative approach is to accept that higher-level knowledge is required to make any early interpretation stage (even thresholding) truly effective and to attempt to make that knowledge available in the form of assumptions about the image(s) being analysed. For a thresholding algorithm, this might include an assumption that the black regions of the input image are, by and large, long and thin. Systems constructed in this way are termed top-down and are, by their very nature, tuned to a particular class of images. If the assumptions upon which they are based hold, top-down systems often produce more reliable results than comparable bottom-up methods. If the assumptions are false, however, performance typically deteriorates very quickly. The need to binarise a wide variety of line drawing images, all with similar characteristics (Chapter 1) has led to considerable interest in top-down thresholding methods, with such methods often being termed direct vectorisation algorithms.

Joseph [81] describes a direct vectorisation method based on a line tracking system which traces over ink marks in the input image, extracting a vector description of the linework without producing any intermediate representations. It is argued [81] that this approach leads to a system which may be applied to images of any reasonable resolution depicting any of a wide range of line drawing types. More importantly, Joseph points out that the tracking paradigm also allows low-level line extraction and high-level interpretation to be integrated in a natural and flexible manner. As Joseph's tracker formed the basis of the ANON system for high-level interpretation of images of mechanical drawings, discussed in detail in Chapter 11, the control/interpretation strategies it employs will be considered later. For the present we shall concentrate on the manner in which Joseph's method binarises an input image.

Joseph's system comprises three main components: a line tracker, a vector store and a book keeping module. The book keeping module initiates tracking and terminates it when all significant lines have been found. To this end the image is divided into a number of sub-regions and the method of Dunn and Joseph [64] used both to produce a normalised image (see Section 4.3.1) and to estimate the number of black pixels in each region. This latter figure is used to direct the line tracker to regions containing large numbers of unexplained black pixels. As each vector is extracted, estimates of its length and width are used to reduce the black pixel count of each region through which the vector passes. Tracking ceases in a particular region when the number of unexplained black pixels therein falls below a threshold level. Having identified a particular region as worthy of (further) exploration, the book keeping module generates pseudo random starting locations within that region and passes them to the tracking module.

Upon receipt of a start location, the tracker begins by performing a circular search around that point. The radius of this search is initially fixed, but later varies as a function of the average width of the lines extracted. Each search generates a circular track of pixels taken from the (normalised) grey level image. A search is then performed for black spots along that track. If only one such spot is found the circular search is repeated with the circle centred on the identified spot. Should two spots be detected, a line between them is hypothesised and linear tracking commences. If more than two spots are discovered the largest two are used to hypothesise a line and tracking again begins.

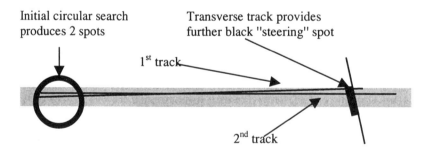

Figure 7.10. Overview of the operation of Joseph's line tracker.

The basic operation of Joseph's line tracker is to test a linear sequence of pixels and, while they satisfy some criterion of blackness, extend that sequence. Given a line hypothesis, i.e. the locations of two black spots, the tracker first attempts to grow a linear track of black pixels between those spots. If this attempt fails, the search is terminated and control is returned to the book keeping module. If the initial tracking step is successful, however, the tracker enters a loop within which it alternately extends its current hypothesis as far as possible and then adjusts its direction to steer closer to the centre of the underlying line. Steering is achieved by examining a second linear pixel track, orthogonal to the developing vector, and using the mid-point of a black spot detected along that track to estimate the vector's centreline. Figure 7.10 shows this process diagrammatically. Note that after each steering step the system attempts to track the pixel sequence from the rear of the two initial spots all the way through to the steering spot; this ensures that any output vectors are fully supported by a chain of sufficiently black pixels.

When no further extension to a given track is possible, a vector description, based upon the pixel track followed and width measurements made as a by-product of the initial and steering searches, is added to a developing line storage module. This records the results of the vectorisation in a manner suited to later interpretation: lines are grouped according to orientation and co-linear lines are stored as such. Joseph's use of a separate line storage module has the advantage that the tracker can find lines in any order, safe in the knowledge that the line storage module will produce an equivalent representation. The particular storage method chosen [81], has the further advantage that it is compact and well organised, allowing easy access to the extracted lines.

Of primary interest at present are the thresholds employed during the tracking procedure. There are several, all based on the image assessment measures proposed by Dunn and Joseph [64]. In the circular search, spot detection is initially achieved using the whitest threshold possible in that region without incurring excessive noise [81]: $g_m - 2\sigma$ (see above). This is later adapted to half the measured blackness of the lines found, as long as this modified value does not fall below the initial estimate. During tracking of linear pixel sequences two thresholds are employed. Meeting a single pixel with a value below a fatal threshold of around 1/3 to 1/4 of the line's greyness stops tracking immediately. A second, blacker threshold (around 1/2 of the line's greyness) is also used to mark provisional line ends. After marking a provisional end, tracking, however, continues and if the pixel track becomes darker again the provisional end is removed. This mechanism, similar to Canny's [166] thresholding with hysteresis, allows small light sections to be jumped as long as further sufficiently dark pixels follow on the same linear path.

Joseph's approach has several strengths. First, the system only examines pixels on or near a hypothesised line. As a result, the computations performed are both simple and fast. Second, because the tracker focuses on one line at a time; it is naturally insensitive to the distractions posed by junctions and other intersecting linework. One might reasonably argue that the detection of junctions is an important part of the drawing interpretation task. Not all junctions, however, reflect the underlying structure of the drawing; many are accidental. The identification of which intersections are important and which are not should, Joseph would argue, be left to a higher-level interpretation system.

Joseph has tested his proposed technique on a number of real and artificial images. Figure 7.11 shows one of the poorer quality drawings with which he was concerned. The result of thresholding this image using the method of [64] at $g_m - 3.0\sigma$ is given as figure 7.12. While the basic structure of the drawing is visible, the binary image is heavily fragmented. The output of the tracking system is shown in figure 7.13. Although the vectorisation is far from perfect, this shows a significant improvement in line identification and neatly demonstrates the value of a top-down approach.

136   Machine Interpretation of Line Drawing Images

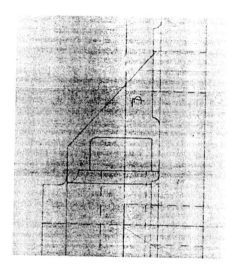

Figure 7.11. Reprinted from Pattern Recognition, 22, Joseph S.H., Processing of line drawings for automatic input to CAD, pp 6, Copyright (1989), with permission from Elsevier Science

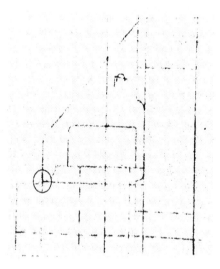

Figure 7.12. Reprinted from Pattern Recognition, 22, Joseph S.H., Processing of line drawings for automatic input to CAD, pp 7, Copyright (1989), with permission from Elsevier Science

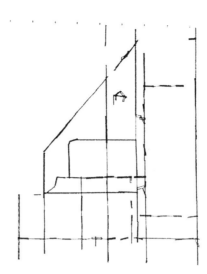

Figure 7.13. Reprinted from Pattern Recognition, 22, Joseph S.H., Processing of line drawings for automatic input to CAD, pp 8, Copyright (1989), with permission from Elsevier Science.

7.6 A Vector Database

The basic idea of a vector representation was discussed in Chapter 2. Having now reviewed scanning, binarisation and raster to vector technologies, however, it is useful to revisit vector descriptions and consider in more detail the vector database developed in [45,122]. This was originally developed to represent automatically vectorised maps, although it has since been applied to other document types including engineering drawings.

If a wide range of drawing objects is to be recognised a structural approach, in which complex entities are represented in terms of their sub-parts, is often employed. At the vector level, this leads to a description of the line drawing in terms of vector segments, their individual properties and interrelations. Superfluous information, however, simply wastes computational resources; a compromise must be found between the richness and physical size of the descriptions produced.

We use a three-level representation that describes both image and vectors. The first and lowest level contains information about the connected components making up the image; the second describes the segments associated with these image regions; the third is concerned with feature points and their arrangement. Let us consider each level in more detail.

The connected components described at the first level are represented by their geometrical and structural parameters. The geometrical parameters recorded are:

- co-ordinates of a minimal box parallel to an axis through the connected component;
- perimeter length (the sum of the lengths of all the vector segments associated with the component's boundary).

These can be used to separate text and graphics. The structural parameters recorded are simply the number of internal contours and feature points.

The segments described at the second level may be either open (linking end points and/or nodes) or closed. The segment description includes the thickness, length, and type of each segment. Segment type comprises a description of the start- and end-point types. Metric information is also stored. This includes the co-ordinates of a set of points lying on an approximating line segment and supports the construction of higher-level geometric entities.

The third level contains a description of any feature points found. This level was introduced because an object's topology is often at least as important as its geometry. At this level we separate descriptions of nodes and end points. Nodes are used to describe connections between segments of the same connected component. Each node description includes the quantity, point co-ordinates and numeric identifiers of segments meeting at this point. End points are used to describe the disposition of segments belonging to different connected components. This is necessary for scene analysis and recognition of complex objects. To reduce the complexity of, and processing time required for, segment analysis, the set of end points is separated into, and indexed via, rectangular image subregions.

Physical realisation of this representation is achieved using three files [122]. The first contains segment and node descriptions and is stored as a direct access file with fixed length records (figure 7.14.a). Each record is separated into three parts corresponding to the three description levels (figure 7.14.b). All the segments belonging to a given component are grouped together by a uni-directional closed list structure. By analysing this list, component characteristics can easily be computed. The third level (information about feature points) is recorded in the last part of the record. This is divided into two subparts, each storing the feature point found at one end of the segment. Either a node number (if the segment is bounded by a node) or a pointer to the next end point in the same subregion (if the segment is bounded by an end point) is stored. As nodes are important for object recognition, their descriptions are stored at a special position within the file; after the last segment description. The second file contains metric descriptions in the form of segment point co-ordinates (figure 7.14.c). The connection between the two files is made by references stored in the first file's records: separate storage of these two files is convenient for segment analysis. In the third file, a 2D array contains the headers of the point sets belonging to each sub-region (figure 7.14.d). An example of the use of this data structure to describe the simple map of figure 7.14.e is shown in figure 7.14.

Vectorisation

t1	s1	e1	e2
t2	s2	e2	n1
t2	s3	e4	n1
t2	s4	n1	n2
t2	s5	e5	n2
t2	s6	e6	n2
t3	s7	e7	e8

n1	3	s2	s3	s4
n2	3	s4	s5	s6

a.

Connected Component	Segment	Feature point First point	Feature point Last point

b.

(x11,y11) (x12,y12)
....
(x71,y71) (x72,y72)

c.

s1,s2	s2,s3,s4
s7,s6	s4,s5,s6

d.

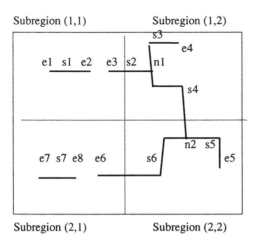

e.

Figure 7.14. Vector representation of a simple map. See text for details.

7.7 Removing Noise from the Vector Model

Like the skeleton and contour data discussed in Chapter 6, vector output is usually subject to noise and distortions that must be identified and corrected once the vectorisation process has terminated. Two types of error correction may be required. The first is concerned with the logical structure of the database. A set of automatic procedures analyses the correctness of the output data structure, checking the existence of all the necessary records, descriptors, references, etc. and corrects any disturbances automatically. The second is concerned with the metric properties of the vector representation.

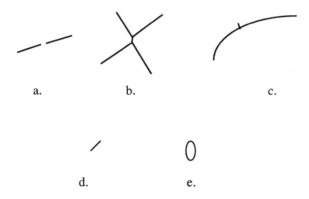

Figure 7.15. Object defect types. See text for details.

The errors found in vector output mirror those seen after skeletonisation or contouring (Chapter 6). Small gaps appear between the end points of successive vectors (figure 7.15.a), implausibly short vectors connect or emanate from nodes (figure 7.15.b,c), isolated short vectors (figure 7.15.d) and vector loops (figure 7.15.e) appear in background regions.

A defect reduction algorithm based upon the analysis of feature points and segment parameters in a vector representation is described in [122]. This deletes defects in the order listed above. The primary operations of the algorithm are deletion and connection of vector segments and points. Gaps are extracted by analysis of the list of vector end points. If the distance between the nearest end points of two successive vectors is less than a given threshold, both end points are deleted from the list and the two segments are merged together. Short branches attached to nodes and multiply connected nodes are dealt with via analysis of a node list. Each segment emanating from a given node is examined and, if its length is below the threshold, deleted from the segment list. Should this leave a node connecting exactly two segments, the node is also deleted. Similarly, if removal of a short segment would separate two nodes, those nodes are merged together to form a new node. Small loops and isolated segments are also removed by applying length thresholds to a segment, details are given in [122].

The number of defects deleted and the overall success of the process obviously depend upon the parameters (i.e. thresholds) employed. These are provided by the operator, but should reflect the resolution of the image and expected size of the drawing objects. Figure 7.16 shows a section of vectorised engineering drawing before and after defect reduction.

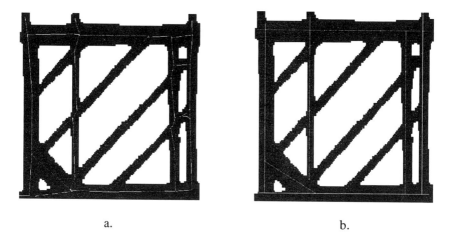

a.   b.

Figure 7.16. A section of a vectorised engineering drawing; a) before and b) after defect reduction.

7.8 Alternative Raster to Vector Technologies

As the topic of this book is *machine* interpretation of line drawing images, we have naturally concentrated on automatic raster to vector conversion methods. It was noted in previous chapters, however, that to date neither map digitisation nor engineering drawing interpretation can be achieved without some human intervention. Existing commercial line drawing digitisation systems typically involve an operator and incorporate various types of interactive raster to vector conversion. Lammerts van Bueren [167] and Janssen and Vossepoel [168] identify the following types of the interactive drawing image interpretation:

1. Blind vectorisation occurs when the drawing to be vectorised is pasted onto a digitising table and the operator indicates the start and end points of each line by clicking on them with a mouse. He/she concentrates on the drawing and has no direct view of the result of the process. This is a fully manual drawing digitisation technology that has been used for some 30 years to input maps and engineering drawings to computers. Its main drawback is that the operator is unaware of any errors made until the whole drawing has been input and a hard copy of the result produced on a suitable plotter.

2. In interactive vectorisation a similar procedure is used; the difference here is that the operator receives immediate feedback as the points and lines entered are displayed on a high resolution computer screen. He/she can see and correct any

errors during the process. The disadvantage of this approach is that the operator must constantly move his/her head between the screen and the digitising table. This may not be desirable from a human engineering point of view.

3. Scanning and heads-up vectorisation requires the drawing to be scanned and displayed in raster format on a high-resolution screen. Vectorisation is achieved manually using a mouse as above. The operator uses a mouse to traverse the screen, marking key points with clicks. The resulting vector data is displayed as an overlay on the original image, providing a better opportunity for error detection than interactive vectorisation without the need for the operator to constantly change viewpoints. Some additional features, such as snapping of mouse-selected points onto a predetermined grid, may be available in this type of system.

The ideal combination of automatic and manual techniques would be true co-operative or interactive intellectual vectorisation. Here, after scanning, drawing interpretation software attempts automatic vectorisation. When the system is not able to deal with the ambiguity present in the image to decide where a particular vector should be placed, the operator is prompted to provide his/her judgement. The software then continues to process the image. This combination of interactive and automatic processing should minimise automatic digitisation errors, decrease digitisation time and make the operator's work at least a little more interesting and less prone to error. We shall return to the topic of interactive interpretation, in the context of map input, in Chapter 8.

Chapter 8
Interpreting Images of Maps

8.1 Introduction

As discussed in our opening chapters, the interpretation of line drawing images can be divided broadly into two stages. A potentially domain-independent raster to vector conversion is usually followed by domain-specific, ideally fully automatic recognition of appropriate drawing entities, specific objects and two-dimensional scenes. In Chapters 4-7 we have considered the various components of the raster to vector transformation. In this and subsequent chapters attention turns to the domain-specific interpretation of line drawing images. Chapters 8 and 9 present our own approach to the interpretation of images of maps. Chapters 10 and 11 focus on engineering drawings.

8.2 System Overview

Much effort has been expended on the raster to vector transformation of map images and many raster to vector systems have been developed. Automatic recognition of cartographic entities, objects and scenes remain, however, research problems. We ourselves have considerable experience of trying to identify cartographic objects and have been quite successful in recognising both line objects (isolines, roads, etc.) and area objects (forests, cities, etc.). It has become apparent, however, that fully automatic recognition of all the entities in a given map is as yet unrealistic. When fully automatic object recognition is attempted it must usually be followed by a significant amount of interactive post-processing to correct recognition errors. A further barrier to widespread industrial acceptance is that most fully automatic systems rely upon powerful workstations or specialised hardware. These facilities may be available to larger companies but if the technology is to achieve its potential, methods are required which can produce high-level representations at reasonable speeds on PC-based systems.

With this background in mind, we propose a combined map interpretation technology that incorporates [169]:

- automatic vectorisation;
- automatic object recognition;
- nteractive object interpretation;

- manual object digitisation.

Automatic recognition and interactive interpretation operate on the output of the automatic vectorisation system. Manual digitisation is an independent process applied to areas of the map that cannot be processed automatically, perhaps as a result of very poor quality or very high complexity. Figure 8.1 illustrates the approach.

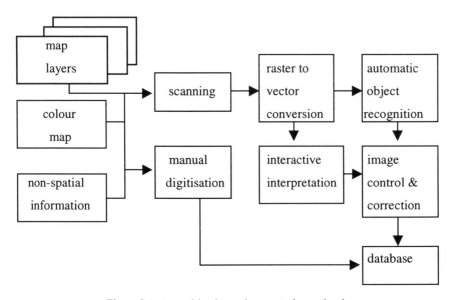

Figure 8.1. A combined map interpretation technology.

We have worked within this framework for some 10 years. In our experience this type of combined automatic/interactive system minimises errors, decreases the time required to digitise a map and reduces the complexity of the operator's task. It allows one to reach an acceptable compromise between digitisation quality, speed and level of automation. It also brings object interpretation under operator control. This makes the operator's actions of primary importance and his/her role much more satisfying than it would be given only the secondary task of correcting recognition errors. Another advantage of this approach is that it supports digitisation of large maps with only limited computing resources. The automatic system does not have to complete the entire task and so can focus on the more manageable areas of the image. Interactive drawing interpretation operations in general do not place heavy requirements on the supporting hardware. In what follows we concentrate primarily on interactive map interpretation systems developed within the above framework; automatic components are the subject of Chapter 9.

Interpreting Images of Maps

8.3 Map Interpretation Principles

In an attempt to provide a generally applicable framework for map interpretation, we suggest that developing map interpretation systems and technologies adopt the principles listed below. Some of these will be familiar from Chapter 2, others are more closely tuned to map interpretation.

1. **Combined automatic/interactive interpretation.** As one can expect only the simplest maps to be successfully digitised by a fully automatic system, a combination of automatic and interactive techniques should be used, following the rough guidelines provided in figure 8.1.

2. **Sequential processing of map layers.** Digital maps are obtained by combining information from separate layers (Chapter 2). If the input to the system includes a map overlay, the order in which the various layers are to be processed should be clearly defined. Imposing a fixed order on the interpretation of these layers allows information obtained from one to be exploited during the interpretation of the next. Experience suggests the most effective interpretation sequence is to process layers containing area objects first, followed by isoline and hydrography layers. The black layer should be processed last. Having interpreted the road layer, for example, information is available which may used to extract road objects and their characteristics from the black layer.

3. **Explicit multi-level recognition.** The recognition problem is naturally (implicitly) multi-level; a feature which should be accepted and made explicit in any recognition system. Object components should be extracted, recognised and relations between them established at the first level. At the second, simple cartographic objects should be recognized and, finally, at the third level, complex objects and scenes should be identified.

4. **Use auxiliary information.** Auxiliary information from other map layers or sources (neighbouring map sheets, neighbouring, or otherwise related, objects) should be exploited wherever possible. In particular, interpretation should start with objects that have associated auxiliary information.

5. **Proceed from the simple to the complex.** In the absence of auxiliary information, interpretation should begin with objects having clear, simple structures and only later move toward more complex entities. In practice this usually means that interpretation should begin with line objects, which are generally simpler than symbols and textured areas and allow one to collect valuable auxiliary information.

6. **Exploit all available knowledge.** Knowledge of map conventions, cartographic objects, scenes, etc. should be used wherever possible. During map production, domain-specific rules and knowledge are used to define the object and scene representations upon which the map is based. During map interpretation this same knowledge can support object recognition. Use of knowledge in the

analysis of complex situations can significantly increase the proportion of objects that are recognised automatically.

7. **Separate knowledge and algorithms.** The interpretation system should be organised so that information and knowledge about objects and map properties is clearly separated from the recognition algorithms employed. Hence if new map types must be interpreted, the knowledge base may have to be updated but, ideally, the recognition algorithms will not be affected. Perfect separation is hard to achieve but even a partial solution could allow significant changes to be made to the scope of the system without large modifications to the system software.

8. **Take a structural approach.** As there are very many different types of map in common use it is not surprising to find that there is as yet no unique recognition method that can be applied to all map images. Our experience suggests, however, that emphasis should be placed on the structure of the drawing during recognition of line objects and symbols from vector data.

8.4 A Classification of Map Entities

If the above principles, particularly those referring to auxiliary information, knowledge and object complexity (3-6, 8) are to be applied, it is necessary first to have a clear picture of the set of entities expected to appear in the given class of map. There are three basic object classes:

- lines;
- symbols, which have restricted geometrical parameters and must be represented by their type and a small number of physical points;
- regions, which should be represented in terms of their bounding contours.

We now consider each entity/object type in more detail.

8.4.1 Line Objects

Analysis of maps and experience of map interpretation suggests that two major types of line object exist and are commonly used, whose manual digitisation is both difficult and expensive. These are isolines and roads. Isolines, sometimes called contour lines, mark strips of land with constant altitude. Road lines denote paths above a certain level of establishment, which varies with map type. Both can be further classified depending on their form, structure and other parameters. Figure 8.2 shows one possible classification of line objects.

At the top level, line objects may be classified as either connected or unconnected. Connected objects are those in which any two points lying on the object are linked by some curve, each point on which also lies on the object. This corresponds to the notion of a connected component of a raster image (discussed in Chapter 5). Unconnected objects are created by arranging connected components

Interpreting Images of Maps 147

into patterns that form meaningful objects within the problem domain. A simple example of an unconnected line object is the dashed line used to represent isolines (contours). This consists of similar line segments separated by similarly sized gaps. By varying the segment and gap parameters (length, width), different types of isoline may be represented. Other unconnected cartographic objects are also formed by repetition of one primitive element (line, arc, etc.). These are often referred to as having a periodic structure or known as one-dimensional texture objects.

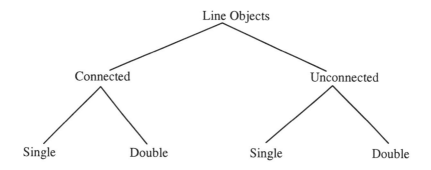

Figure 8.2. A taxonomy of cartographic line objects.

Line objects may be formed from single or double lines. The dashed line mentioned above is a single line object. Double line objects are formed from pairs of closely spaced parallel segments. Double line objects may be connected or unconnected. Examples of cartographic line objects of various types are shown in figure 8.3.

The semantics of line objects vary from country to country. On Russian maps, for example, the unconnected line style shown in figure 8.3.a is used to represent isolines, while the objects shown in figure 8.3.b-d mark the borders of administrative regions. Tree stripes are usually denoted by the unconnected pattern of small circles shown in figure 8.3.e. Figure 8.3.f represents coast banks, 8.3.g cutting in forests, 8.3.h,i,k,l fences, 8.3.j gas or water pipelines (depending on colour) and 8.3.m embankments. The major parameters of line objects are length and thickness of segments, length of gap(s) between segments, distance between segments in double line objects and the structure (e.g. straight, circular, etc.) of the component primitives.

8.4.2 Area Objects

Like cartographic line objects, area objects may also be divided into two groups. The first comprises areas of solid colour (i.e. black), the second comprises areas filled by some more sparse texture. Solid black areas are comparatively simple to process, the most common approach being to extract bounding contours and area parameters (see Chapter 5).

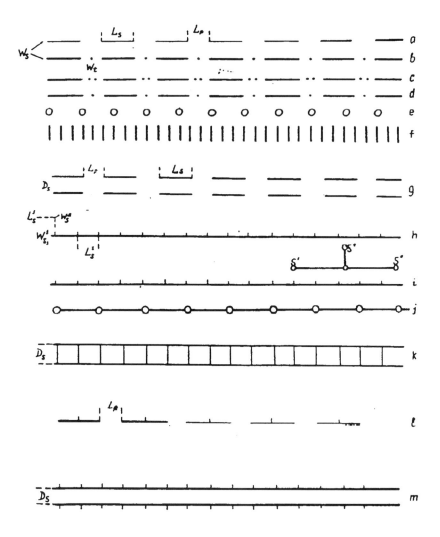

Figure 8.3. Types of line objects. Annotations describe object parameters.

In what follows we concentrate on textured objects. These are typically formed by repetition of some, usually small, solid black primitive called a texture element. While textured areas (e.g. crosshatching) do appear in engineering drawings, they are much more common in map images; gardens, forests, sand, etc. are all represented using texture. The use of textured objects in cartography makes map interpretation much more difficult; indeed most current systems do not attempt to process these types of entity.

Texture analysis has been studied quite intensively, in other image analysis contexts, for some twenty-five years. Most texture analysis techniques have, however, been developed for and so only applied to the natural textures found in grey level (often satellite) images. Most are empirical and seek to define grey level

Interpreting Images of Maps

texture in statistical terms [170,171]. The textures used in cartographic objects differ from those found in grey level images and require special interpretation techniques. The main purpose of texture on a map is to indicate the boundaries of, and to classify areas of land. Since the patterns of repetition of texture elements used in line drawings are generally very regular, map textures usually have strong structures. As a result they are often referred to as structural textures [172]. Natural textures reflect the intrinsic properties of visible surfaces and are often more variable as a result.

A texture element is a group of pixels (objects) characterised by some shared set of features. Textured areas are obtained by structured repetition of texture elements, the amalgamation of the texture elements defining the physical extent of the area. Line drawing textures can be classified using only three primary parameters:

- the size of the texture element;
- the periodicity rule used to guide texture element placement;
- the dimensionality (1- or 2-) of the object.

Two categories of texture objects can be distinguished purely on the basis of texture element size [172]:

- <u>Macrotexture</u> arises when the texture elements employed are sufficiently large that the human eye can estimate the shape and the size of each texture element on the original line drawing. Examples of the use of macrotexture in cartography would be a swamp (figure 8.4) or garden on an ordance survey map.

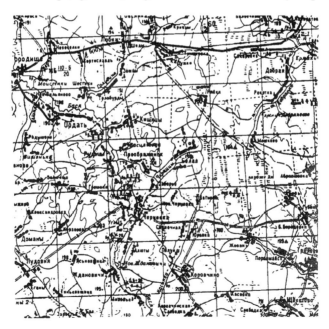

Figure 8.4. A topographic map containing macrotexture.

- Microtexture arises when small points or thin lines are situated sufficiently close to each other that individual texture elements are not distinguished by the human viewer and so can have no independent meaning. The image of a river map layer shown in figure 8.5 provides one example of a microtexture.

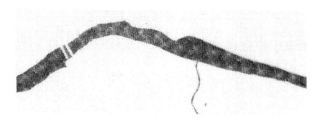

Figure 8.5. Microtexture on a river map layer.

Three types of microtexture are commonly found on line drawings (figure 8.6.):

- point (otherwise dot or blob) textures formed from a set of small, compact black regions;
- line textures comprising thin parallel lines;
- net textures formed from two sets of thin perpendicular lines.

Textured objects may be divided into two classes, regular and irregular, on the basis of the texture element periodicity rule used. Regular textures are produced by the repetition of the same texture element according to a strong periodicity rule. The point, line and net textures shown in figure 8.6 are clear examples of regular textures. Irregular texture arises when the texture elements involved vary in size and orientation (for example, sand on a map). Textured objects may be one-dimensional (i.e. lines with repeating structures) or two-dimensional (i.e. texture regions).

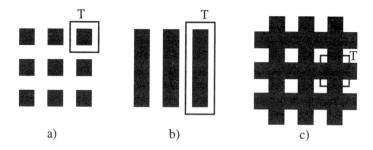

Figure 8.6. Sample texture types; a) a point texture, b) a linear texture and c) a net texture. Individual texture elements are bounded by rectangles and labelled T.

8.4.3 Symbols

The final group of cartographic objects is quite wide-ranging and includes general and domain-specific symbols, characters and textual information areas. Symbols can also be divided into two major groups: natural language characters and cartographic signs. The characters on a map are obviously the same as any others, though they can be situated pretty much anywhere on the map in any orientation and at a fairly wide range of scales. Standard characters and text strings may, however, be constrained to appear in certain special relationships to other cartographic signs. There is rarely any meaning in a single character; words must be recognised and any connections established between them and neighbouring, corresponding, cartographic signs.

Around 500 different cartographic signs appear in Russian maps, with a similar number being used in most other countries. This is obviously much more than the 30-40 characters and digits recognised by most OCR systems. Moreover, cartographic signs usually have a much more complex structure than simple alphanumeric characters. They also tend to appear as groups; to recover all the available information about a labelled object, each of these groups must be extracted and interpreted as a unit (figure 8.7). Cartographic symbol recognition is a complicated problem and not yet fully solved [173-175].

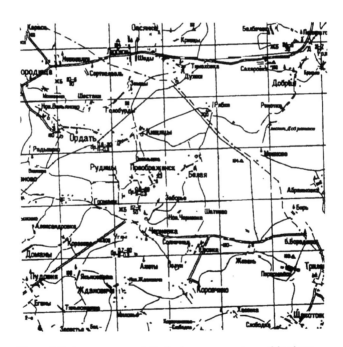

Figure 8.7. A map segment displaying many cartographic signs.

8.5 Interactive Map Interpretation

Having described our proposed framework for combined automatic/interactive interpretation of images of maps, discussed the underlying principles and considered the drawing entities which must be processed, we now present two different map interpretation approaches based on our model.

8.5.1 Automated Image Interpretation under Operator Control

This digitisation mode relies upon a modified, automatic entity recognition system operating under human control. It is particularly well suited to the recovery of line objects. One of the most difficult tasks in map interpretation is the identification of the long, twisting line objects that appear as extended strings of line segments in the vector database. To simplify the digitisation of such entities, which mainly represent roads and isolines, automatic recognition is performed under the visual control of a human operator. The process comprises two tasks:

- **Segment analysis,** in which vectors are grouped together via a tracking procedure which decides whether or not to extend the developing object after matching the pattern of lines found to a template or "etalon" description of the entity at hand. The algorithms employed are similar to those used in fully automatic recognition (Chapter 9) though the conditions imposed on vector grouping are somewhat less strict.

- **Interaction** with the operator when situations arise in which the automatic system cannot identify a unique solution to some problem.

A major drawback of traditional manual input methods is the difficulty of providing visual control to the digitisation process. The user typically finds it very hard to tell which objects have been digitised and which are awaiting input. In our system this problem is overcome by simply displaying raw (i.e. unprocessed) data on the screen in dim colours and using bright colours to denote entities that have already been digitised. The user is provided with a range of input tools (the original image in raster form providing the background, microscope, moving glass, changing pick square, etc.), all implemented in software, and can select those that make his/her work more comfortable.

Manual digitisation of elongated, twisting line objects is usually achieved via a follow-glass technique. In a system based on automatically vectorised data there are opportunities to ease this process. The vector representation contains the co-ordinates of points lying on these objects: to identify elongated objects in a vector database the user need only indicate component segments (rather than points) in some appropriate order, perhaps by selecting them with a mouse. This simplification of the digitisation task makes map input both faster and less wearisome.

Many line objects can be identified almost entirely automatically. For example, consider the algorithm used in [176] to digitise dashed lines. This differs from previous approaches [158,177,178] by using a deeper analysis of the expected situations to guide the grouping of line segments. Output is noticeably more reliable

Interpreting Images of Maps   153

as a result. The technique was developed specifically to deal with images of complex maps in which various forms of dashed line are to be found, many with their component segments situated very close to each other.

The algorithm [176] relies on a set of dashed line etalons (or templates). The etalon for each type of dashed line includes:

- minimal and maximal width and length of component segments;
- minimal and maximal distance between neighbouring segments;
- minimal number of segments in the line.

Each input vector that matches an etalon description is considered in turn. Having picked an initial segment and identified a potential etalon, the algorithm tries to assemble a segment chain of the corresponding type. The process is iterative. During each iteration the last segment of the chain is examined and an attempt made to connect its free point (the chain tail) to some other segment. If this attempt is successful, the newly added segment becomes the last in the chain and the operation is repeated. When the chain cannot be extended any further in the initial direction a similar attempt is made to continue the opposite end. If the final dashed line does not contain sufficient segments it is discarded and the algorithm tries to assemble a different type of dashed line around the initial segment. If none of the available etalons provides an acceptable dashed line, attention moves to the next unconsidered line segment

Throughout the process the developing dashed line is characterised by a description, associated with its tail, which contains both the co-ordinates of the chain's end-points and a vector giving the orientation of a straight line between them. When extending the chain, we consider a local neighbourhood of the appropriate end-point. If this is a node (Chapter 5), all the segments that meet there are considered. If it is an end-point, we identify and consider all segments lying within a neighbourhood whose size is defined by the current etalon's maximal gap. Segments whose parameters (length, width, etc.) correspond to the given etalon are marked and considered for inclusion in the chain. Distance D from the chain tail and connection angle A (figure 8.8) are calculated for each candidate. The connection angle is defined as A=min(a1,a2) if the connection is made through a gap but is simply the angle between the two adjacent segments when they meet at a point (figure 8.8).

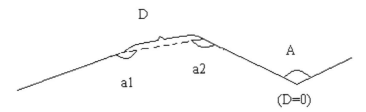

Figure 8.8. The parameters of potential continuations.

The space of all D and A values is divided into zones (figure 8.9), each of which is given both a number and a descriptive, symbolic name (for example, "very near and straight ahead", "not far and not straight ahead", etc.). Zone numbers are used to rank the set of potential continuations: more preferable zones have smaller numbers. If two potential continuations have the same zone number, the nearer segment is considered the better. When continuations have equal D values, the one with the smallest connection angle is selected.

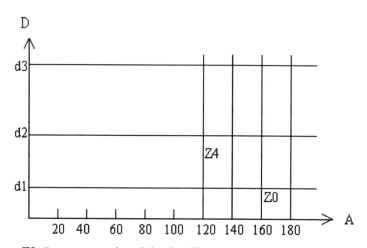

Z0: "very near and straight ahead"
Z4: "not far and not straight ahead"

Figure 8.9. Continuation zones.

Once the most likely continuation has been found an attempt is made to either confirm or refute it. This is done by applying the continuation process to the near end of the proposed new segment. The effect is to check that, were the dashed line to be traversed in the opposite direction, the algorithm would again consider the proposed connection to be the optimal one. The continuation is confirmed if, when applied in reverse, the extension process considers the current end of the chain to be the best next segment. When this occurs the proposed continuation becomes the new chain tail. If, however, a different solution is proposed by the confirmation step, an attempt is made to confirm or refute this variant in the same way. This process continues until a pair of segments is found that cannot be refuted. This pair is marked to exclude them from further analysis and the system tries once again to extend the current chain. The extension process terminates completely when no suitable, unmarked segments are available.

This type of automatic recognition technique (see also Chapter 9) has been incorporated into our interactive map interpretation technology. For example, if the operator is confident of the correct continuation of a developing chain line, he/she need only confirm the decision of the system. Several forms of elongated object digitisation have been realised within our system. For simplicity we refer to them as the Draw, Pick, Go, Run and Jump modes.

Interpreting Images of Maps 155

- Draw mode allows the user to input object co-ordinates via a mouse and is mainly used to digitise text, symbols, poorly scanned lines and regions. The user can work over a raster image background that provides contextual information to aid the digitisation process.

- Pick mode allows the operator to choose vector segments to be connected, again using a mouse. The system automatically selects the manner of connection between the developing chain and each newly picked segment and proposes this form of continuation to the user. The nature of the proposed connection is determined after consideration of the relationships between segments already comprising the chain and the distance between the chain end and the new segment.

- In Go mode, the system automatically extracts possible continuations, ranks them using the criteria employed in Pick mode and proposes the best to the user. Some five or so ranked options are usually provided. The operator may select one of these or, if his/her preferred solution is not among the list, view other variants or switch to Pick mode to indicate the required segment.

- In Run mode, the system tries automatically to join segments to form a chain of some predetermined length (around 15 segments are typically combined at any one time). The connectivity criteria used in Pick mode are again employed. The difference between the connectivity score for the best and next best solutions provides a measure of confidence in each continuation. If the confidence of a given continuation is high the system will proceed to attempt the next extension; otherwise it switches into Go mode and asks the operator for confirmation.

- Jump mode is a combination of Pick and Run: the user specifies a start segment and an end segment which the system then tries to connect with a high confidence chain line. The solution is presented to the user for confirmation. To this end a tree of possible continuations is formed. Every possible continuation has an associated weight and maintains a reference to its parent; the previous segment of the proposed chain. Initially, the tree comprises only one segment having zero weight. To grow the tree a segment with no children and minimal weight is selected. Other vectors, which might form acceptable continuations to the selected line, are then identified and added to the tree as children of the current segment. The weight of each new child is calculated as a linear function of the weights of the segments in the chain, the length of the new segment and its distance from both its parent and the end segment. The procedure terminates when the user-defined end segment is added to the tree. Backtracking through the tree to the start node then produces a chain line. The method has proved effective for objects that have many connected segments and not many gaps. An example of tracing in Jump mode is given in figures 8.10 and 8.11.

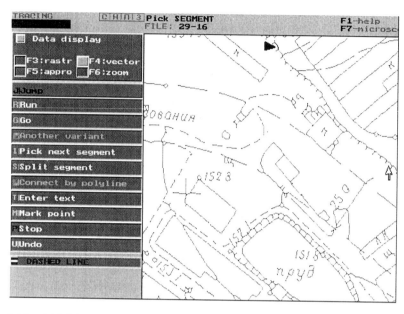

Figure 8.10. Initialising Jump mode extraction of a chain line. The current chain tail is marked with a triangle; the end segment is pointed to by an arrow cursor.

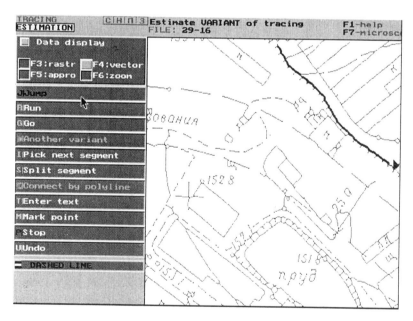

Figure 8.11. The result of the Jump operation. The new chain tail is marked with a triangle, the proposed chain is denoted by a thick line.

Throughout interpretation, all the system's actions are performed under operator control and their results displayed on a monitor. The operator can change the system's decisions at any time and easily correct its results as problems arise, a task much less onerous than post hoc examination of a complete interpretation. Moreover, the above system typically produces results comparable to manual digitisation in around one third of the time required for completely manual input.

Although the system described here is designed for use on topographic maps, the basic methodology is extendable to other kinds of line drawing. The technology has been tested on cadastral and special forest maps with good initial results. Evidently, some elements of this approach are relevant to the wider line drawing interpretation problem.

8.5.2 Operator-Supplied Context

This approach to interactive map interpretation is based upon the idea of combining automatically acquired vector data with manually input contextual information. The map is first scanned and vectorised. Object type, characteristics and approximate object co-ordinates are then input by the operator. This data is integrated and the coarse object co-ordinates refined to produce an accurate entity description.

In the current system, the raster image and vector data are presented to the user via a graphics monitor. Vectors form an active (i.e. accessible) layer on top of the fixed raster image background. An image of the monitor screen is given in figure 8.12. The operator decides which object is to be extracted, then chooses and inputs an appropriate code for that object. He/she then selects, using a mouse, the vectors from which the object is to be constructed. The system extracts rough object co-ordinates from the selected vectors and computes the parameters of the target object. The result is an object-oriented data structure containing object type, geometric co-ordinates and other characteristics, including relations to other objects.

The user may, if desired, provide rough object co-ordinates directly. He/she may describe the target entity using any of three data structures: a labelled point, a labelled region, or a labelled set. A labelled point specifies the co-ordinates and characteristics of one marked point and is used in the digitisation of cartographic symbols. A labelled region specifies the boundary of a region containing objects with similar characteristics. A labelled set comprises the coordinates and characteristics of a variety of different objects. Figure 8.13 shows the user interface to the interactive digitisation system, the set of functions provided to support digitisation of line objects are shown in the bottom left corner of the menu.

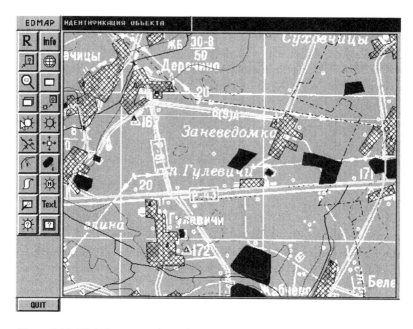

Figure 8.12. Digital map creation using a scanned map image as background.

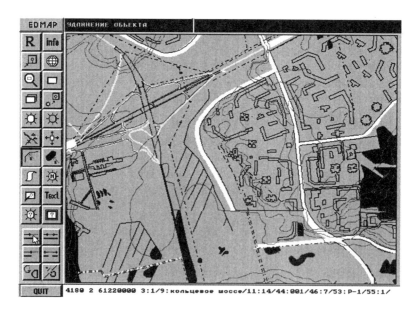

Figure 8.13. Interactive digitisation of line objects.

The two interactive digitisation modes described above have been developed as part of our map interpretation system. The combination of these modes with automatic vectorisation and recognition has allowed us to input and describe all the cartographic entities found on topographic maps. The combined automatic/interactive approach has been widely tested and successfully applied within the (Russian) cartography community. Moreover, the proposed framework provides a robust development path: as reliable automatic interpretation methods become available they can easily be incorporated into the system, perhaps replacing manual/interactive components.

8.6 Output Formats

Once interpretation is complete the digitised map must be described in some appropriate output format and recorded in a database. Many formats and standards have been used to represent the spatial information obtained from line drawings. The earliest were highly task specific, designed for concrete problems and restricted application areas. Examples include the American DEM, DLG, GBF/DIME, DX-90 and TIGER formats [179,180]. During the 1980s more universal formats such as SDTS, SAIF, and NTF became available. These are characterised by their ability to represent and transmit any type of spatial data without loss of information. The drawback, however, is that they are quite complicated and tend to produce large files.

Perhaps the most famous and commonly used conventions are DXF and ARC/INFO Generate/Ungenerate [181]. Formats such as MIF/MID MapInfo, ArcView SHAPE File, Atlas BNA are also popular. Among the most developed formats is the American standard SDTS; variants of which are in use in other countries, e.g. ASDTS in Australia. SDTS allows one to store most types of spatial data, including vectors and raster images.

VPF (Vector Product Format), a standard of the United States Ministry of Defense, was introduced in 1992. This is oriented toward platform-independent storage of large geographical databases and supports most data types. DCW (Digital Chart of the World) was created using this representation.

The most popular formats for the representation of simple vector data are DXF and ARC/INFO Generate/Ungenerate. Converters from DXF are incorporated into practically all cartographic/GIS systems. If vector attribute data must be represented, however, DBF and CSV should probably be considered.

Whatever the format employed, the results of map interpretation must be made available to interested parties as some form of database. The database developed to record the output of our system [169] employs a relational model and consists of three files: metrics, characteristics and inquiry. Records in the characteristics and metrics files are of variable length; inquiry file records are fixed length. The inquiry file contains an object classification code, relationships to other objects, references to metrics, and semantics. The metrics file contains geometrical object coordinates. The characteristics file contains information about object characteristics that are given in accordance with the Russian system of classification and coding of cartographic information. The associated database management system provides the following functions:

- access to information about maps, layers and/or map regions;

- access to information about specific objects or objects with particular properties;
- access to objects by using coordinates and semantic properties;
- transformation of maps, including merging of map sheets, splitting, zooming, projection, etc.

8.7 Quality Issues

8.7.1 Basic Concepts

It should be clear at this point that map interpretation is a complex, multi-stage process in which drawing complexity, defects, distortions and noise conspire to disrupt any system output. Even minor errors in the interpretation of a map can cause significant problems for later processes. Various aspects of error extraction and correction in Geographic Information Systems have been considered in the literature [182-184]. Each digitised line drawing must be checked and its quality assessed. As different levels of interpretation are appropriate to different tasks (Chapter 2), data quality must be assured at/after each stage in the process. We now consider the basic notion of the quality of digitised map data.

The quality of a digitised map or line drawing is defined by a set of common and specific features and properties that measure the data's ability to satisfy certain domain-specific requirements. Quality assurance methodologies provide a set of related principles, methods, algorithms and procedures which check, improve and allow the recovery of data quality should that data become corrupted [185].

Data quality in GIS (or CAD) systems is defined by the quality of both the initial line drawing and the data produced at each stage of the interpretation process. Data quality at the different interpretation stages is characterised by properties such as:

- validity;
- distortion;
- redundancy;
- compaction;
- integrity.

Data validity is the most important component of data quality. Data validity simply implies that there are no logical errors, i.e. that any statements made about and labels applied to data sets are logically correct. Until the entity recognition stage we cannot say anything satisfactory about data validity. In the early stages data quality is really only defined in terms of data distortion. This refers to the geometric correspondence between, say, a vector description and the region of black pixels from which it arose [185]. Redundancy is hopefully self-explanatory, referring to the absence of unnecessary repetition in the digitised drawing, while compaction requires the data to be represented accurately and completely in a small volume. Integrity implies that the data is valid, self-protected against distortion and suitable for use in the domain for which it was intended.

Data quality is also characterised by secondary properties such as accuracy, resolution, accessibility, testability, recoverability and fault-tolerance (see below).

Interpreting Images of Maps

During interpretation these properties promote a gradual shift in focus from distortion, redundancy and error to validity, compaction and integrity.

The task of ensuring data quality in a GIS (or CAD) system is complex; a combination of automatic and interactive tools must generally be used to produce data with the desired properties. The first step is the data quality check. This is followed by the application of data manipulation techniques that attempt to improve quality and recover from distortions. Automatic recovery methods often can only deal with a small number of very precisely defined situations. The operator usually has to assess the changes needed by eye and make them manually via some form of interactive editing system.

8.7.2 Maintaining Quality during Interpretation

We now consider each stage of a typical interpretation process in turn and discuss the application of the above notions to the operations involved [186].

After binarisation the most important data properties are accuracy, resolution and accessibility. In response to a redundant representation (the complete binary image) and a pressing need for compaction, the binarised image must be represented in a compressed form providing satisfactory access to individual binary pixels. The binary image, though compact, will still suffer from noise and other defects (Chapter 5). Many of these can be corrected, due to their properties of testability and recoverability. However, as a rule, all the defects cannot be eliminated from a binary image and some will go on to influence subsequent processes.

During the raster to vector transformation properties such as accuracy, accessibility and compaction become more important. To provide the required data quality as vectors become entities, attention must focus on properties such as accuracy, cumulativeness and accessibility. After automatic entity extraction errors might include incorrect entity type, metric and/or semantic information. Relationships between objects and their components might also be wrongly assigned or labelled. Data validity must be checked at this stage by comparing entity type and parameters with corresponding characteristics in a reference model. Recovery of valid data may be performed by further automatic entity extraction or via interactive editing. During manual editing cumulativeness, testability, recoverability and fault-tolerance are the primary measures of data quality. The primary data test at this stage is a simple sight check; does the interpretation look right to a human operator?

When the interpretation process is complete, the resulting representation must be subjected to a number of checks. It is generally necessary to examine:

- the logical correctness of the data structure;
- any metric descriptions;
- any semantic object descriptions.

Logical correctness is usually ensured by consistency checks built into the database. Examination of metric object descriptions begins with a straightforward visual check of completeness and correctness performed by comparing the final data structure with a copy of the original map. Some automatic checks may also be made based upon cartographic object extraction rules stored in a knowledge base. Semantic tests include checking for the presence of required object characteristics, examining the values of object parameters, ensuring correctness of object codes, etc.

Each of the checks outlined above may be performed either automatically (though this is not always possible) or by producing and applying a control protocol, effectively a list of checks to be carried out by an operator. Software tools may be used to aid the application of a control protocol; these should support the following functions:

- visualisation of the required map area, providing a choice of display device, window, zoom, colour map, etc. and allowing selection and display of objects with given characteristics, codes, etc.;
- editing metric object information, including searching for objects with given co-ordinates, translating and rotating objects, splitting and merging objects, adding and deleting object components and the input of modified object parameters;
- editing semantic object information, including changing object codes, adding and deleting object characteristics, etc.;
- editing logical information (adding and deleting objects, co-ordinating metric object descriptions, changing object relations, etc.;
- parameter calculation (length, area, perimeter, etc.).

Having considered the entities to be recovered from map images, presented and discussed a framework for combined automatic/interactive interpretation of those images and reviewed output formats and quality issues, we now describe the automatic components of our system in more detail.

Chapter 9
Recognising Cartographic Objects

9.1 Recognising Isolines

9.1.1 Overview

The recognition of dashed, chained and other broken lines is one of the classic problems in line drawing interpretation; many broken-line extraction algorithms and frameworks have been developed in recent years. The most common approach is to thin a binary image and vectorise the resulting skeleton [17,157]. The approach produces representations that are both closely related to the required output format and well suited to approximation and/or structural/syntactical line recognition. Among its drawbacks are that it is very sensitive to small irregularities in the line's contour; small branches are produced which must be pruned away afterwards, junctions also may not be properly represented without some noise reduction [122].

A number of techniques for dashed line recognition are available. The algorithm proposed in [158], for example, allows dashed lines to comprise either single length segments or alternating long and short lines. A two-stage algorithm to detect dashed lines is described in [177]. Line segments are first grouped together using local constraints. Context-dependent, global syntax rules are then applied to resolve any conflicts in interpretation. An approach based on a similarly global view of the drawing is described in [187]. Here the input image is divided into rectangular regions and intersections are sought between drawing lines and region borders. An effective sparse-pixel algorithm for bar recognition is proposed in [188]. This also performs a careful sampling of the image before focusing attention on areas identified as key. The Hough transform (Chapter 7) provides a useful general method of grouping together broken line segments. Any construct hypothesised by a Hough transform must, however, be checked for consistency in both segment length(s) and inter-segment gaps.

To provide a forum for comparison and promote the development of improved broken-line extraction techniques, graphics recognition contests have been held at the biennial International Workshop on Graphics Recognition since 1995 [189,190]. In 1995 the contest was limited to dashed-line detection. The rules required automatic extraction of broken lines from test drawings at three levels of difficulty; simple, medium and complex. These contained a variety of straight and curved, dashed and dot-dashed lines, some incorporating interwoven text. Performance evaluation methods, as well as descriptions of the test images, are given in [189,191].

The techniques developed by the winning team of Israeli researchers are described in [189]. A similar dashed line extraction contest was run during the 1997 workshop [190]. Perhaps the most important question raised by this type of competition is what makes one graphics recognition algorithm better than another. We shall return to the issue of performance evaluation later.

The algorithm described here differs from previous approaches by using a slightly deeper analysis of the situations likely to arise in a map to guide the combination of line segments; the result being a marked increase in reliability. The technique was specially developed to deal with the complex situations that arise within map images when contour lines interact with other entities and individual segments are situated very close to each other.

9.1.2 Theoretical Background

In our experience, structural pattern recognition generally provides the most robust approach to the identification of complex line objects. Structural pattern recognition relies on two processes. A description of the object(s) of interest must first be formed. This description is then used to classify input data, recognising instances of the described pattern(s). To represent the dashed and other line entities with which we are concerned, we adopt a method based on formal grammars [193]. The grammatical approach is based on four key notions:

- terminal elements;
- non-terminal elements;
- starting elements;
- grammatical rules.

Terminal elements are the primitive objects from which more complex entities are formed. The terminal elements used by the isoline grammar are the line segment (s), gap (g), point (p), and closed contour (c). We divide terminal elements into two groups: constructive and connective. Constructive elements are those that determine the geometric properties of the line object, connective elements are used to link constructive elements together. Similar notions are used in the ANON system (Chapter 11). Each terminal element has a number of associated attributes (length, width, type, etc.) which carry detailed information about the isoline and its components.

Descriptions of more complex line objects are obtained by successive combination of terminal elements. It is often useful to identify and assign names to those groups of terminal elements that appear in a number of line objects. The ability to make explicit reference to these non-terminal elements both eases the line description process and reduces the size of the descriptions produced. The ways in which terminal and non-terminal elements may be created and combined to form increasing higher-level constructs are specified by grammatical rules. This combination of elements clearly must begin in some well-defined initial situation; to this end one or more objects are labelled starting elements. Entity extraction must commence with the identification of a starting element. Starting elements are naturally terminal.

Isolines consisting of a basic repeating pattern of open line segments and gaps are described by the rule

$P : G \dashrightarrow sgG$

while those comprising open line segments, points and gaps are described by the rule:

$P : G \dashrightarrow sgpgG$

where G is a non-terminal element. In their pure form, terminal and non-terminal elements are simple symbols. In our system terminal elements are data structures storing the following attributes of the primitives they represent:

$s =::$<thickness><type><length>;
$g =::$<length>;
$p =::$<diameter>.

Non-terminal elements are increasingly complex data structures formed by combining these basic descriptions. Examples of the description of other line objects can be found in [193].

In addition to the grammar, which specifies the types of relationships that must exist between terminal and non-terminal elements if they are to constitute a particular drawing entity, the algorithm also employs a set of dashed line templates [193]. The template for each line type comprises:

- the minimum and maximum width and length of component line segments;
- the minimum and maximum distance between neighbouring segments;
- the minimum number of segments required for an instance of the template to be accepted.

In effect, the templates impose constraints on the various parameters of a dashed line's component segments and the relationships between them: the grammar makes sure the pattern of lines is qualitatively correct, the templates ensure they are quantitatively so.

9.1.3 An Isoline Recognition Procedure

To extract isolines from a vector representation of a map, we consider each line segment that corresponds to some template description and, using that segment as a starting element in the grammar, attempt to assemble a chain of segments of the appropriate type. Starting from a single initial segment we search for other segments which, according to the grammar and template, may be used to extend the developing isoline. During each (attempted) extension, the most recently identified segment of the chain is analysed and effort is made to connect its free point (the chain tail) to some other segment. If this attempt is successful, the newly added segment replaces the chain tail and the process is repeated until either an acceptable line termination pattern is found or continuation becomes impossible. In the latter case, the chain is reversed and an attempt is made to continue it in the opposite direction. If the number of segments comprising the final isoline is insufficient, the chain is discarded. Attention then returns to the initial segment and attempts are made to assemble any other line types whose templates match the initial segment.

At each stage in the above process the developing isoline is characterised by a description, associated with its tail, which contains its end-point co-ordinates and orientation with respect to the horizontal. When seeking a continuation, the algorithm examines a local neighbourhood of the corresponding end point, looking for vectors that might form part of the current object.

To simplify and speed up the isoline extraction process, the available vectors are first classified on parameters such as segment length, thickness, and type. Thirteen such classes (c.f. figure 8.3) have been found useful in the interpretation of topographic maps. As the starting segment of a given isoline, we choose a line segment bounded by, ideally two but, at least one end point. If both ends of the segment are free, the segment's length must fall within the predetermined limits stored in the appropriate template. If only one end is free, the segment must either be shorter than the upper limit specified or part of a simple graphical primitive whose length satisfies the first requirement.

Isoline assembly begins at an end point. During tracking two basic situations arise and must be resolved:

1. continuation through a gap;
2. continuation through a node.

In the first case a segment of the same class, offset by an amount corresponding to the length of the expected gap, is sought. A square region of interest is defined with side length 2S, where S is the length of the longest gap allowed by the template. Only a part of the square is actually considered, its location and angular extent depending upon the slope of the last segment of the line (angle A in figure 9.1). All segments belonging to the desired class (in the sense of figure 8.3) are considered first. If there are no segments of this class within the search area, segments from other classes are examined. Should there be no segments in the zone, the side length of the square is increased.

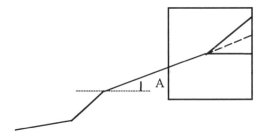

Figure 9.1. Searching for an isoline continuation.

Segments are sought whose end-points are close to the current chain tail and which are collinear with the current segment. All segments lying within the search area are examined and those whose parameters (length, width and distance from the tail) correspond to the current template are considered to be candidates for chain continuation. For each candidate we calculate the distance (D) and angular (A) measures described in section 8.4.1 and illustrated in figure 8.8. As described in Chapter 8, the space of possible A and D values is divided into numbered zones (figure 8.9). The zone number associated with each candidate is used to rank possible extensions; lower zone numbers are preferred, with the nearer candidate being selected when two candidates lie in the same zone. The effect is to choose the segment with the most similar orientation (W1 and W2, figure 9.2). The chosen segment is then added to the developing isoline description and tracking continues.

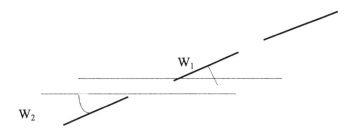

Figure 9.2. Extending a dashed line.

When the current isoline terminates in a node, the algorithm first examines all the vectors meeting at the node and seeks a segment that meets the template requirements of length and orientation. If no acceptable extension presents itself, we assume that we have found a type 1 degenerate case (figure 9.3.a) and attempt to extend the tail line through the discontinuity. Tracking is terminated if we fail to find an extension of the line that satisfies the restrictions listed earlier or if we choose, as an extension, a segment whose vectors make a relatively small angle at the junction. In the latter case a type 2 degenerate case is signalled (figure 9.3.b). When this occurs, an attempt is made to extend the line in the opposite direction.

Figure 9.3. Degeneracy in the recognition of a dot-dashed line; a) type 1 and b) type 2.

During the search for an extension, the criteria for acceptance are relaxed. Several possible extensions are tested, concurrently, to see if they might extend the line through a node, an end point, or a potential degenerate case. At this stage in the proceedings lines may reasonably be extended through discontinuities and nodes, even if the angle at the junction is not quite as it should be.

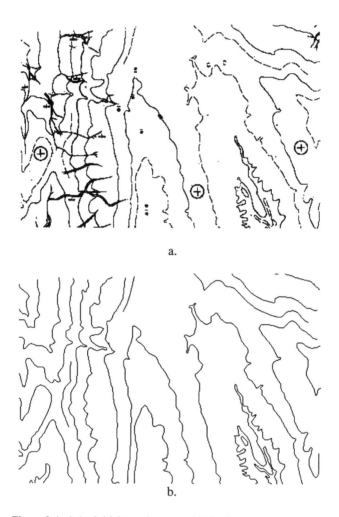

Figure 9.4. a) An initial map image and b) isolines extracted from a).

When searching for an extension through a discontinuity, we examine local nodes as well (using lists of connecting nodes attached to the vectors concerned), so as to detect type 1 degenerate cases. We also analyse carefully any situation in which type 2 degeneracy might occur. All the possible extensions through the discontinuity are ranked on total length of the constituent segments and gaps and the shortest alternative is considered first. A possible extension is considered valid if it involves the end of another isoline or a vector that complies with the template. In this way, we find the shortest path joining the end of isoline to elements included in the template description. The search is terminated if the shortest alternative is longer than permitted by the template. Figure 9.4 shows an example of isoline recognition. It should be noted, however, that some interactive post-processing was required to produce these results.

9.2 Recognising Roads

9.2.1 Overview

The other major line object typically extracted from map images is the road. Knowledge-based interpretation of road map layers is considered in [194,195]. In this work, roads are extracted from an initial raster image via the computation of two sets of distance skeletons: termed the endo- and exo-skeletons. These are derived from a variation of Blum's medial axis transform that preserves both topology and the Euclidean metric. To eliminate noise and quantisation effects, the medial axis is pruned down to its stable inner branches. After the removal of artefacts and further simplification of the skeleton, meaningful structures are identified via a multi-stage procedure with the aid of a growing amount of domain-specific knowledge [194].

A method of extracting road networks from images of urban maps is proposed in [196]. The road network is a graph structure representing, as one might expect, connective relationships between roads. An approximation to the road network is first constructed using a bottom-up approach. It is then refined in top-down fashion using both a priori, heuristic knowledge of urban maps and a set of (in)consistency checks operating within the developing network.

The recognition of roads and railways is considered in [197]. Both are expected to appear as long lines of unknown shape but displaying low curvatures and near-constant widths. A line tracking approach is therefore adopted. The tracker first seeks a long continuous line of some given width. The variation in road and railway line properties in real images of real maps means that the resulting line section is usually only a part of a longer entity. In an attempt to complete the extraction, the initial line is extended from both ends using the same tracking algorithm but with dynamically varying parameters. Further road recognition algorithms are described in [198,199].

In map overlays, two layers contain roads. The first, the road layer, comprises mainly wide ("thick") roads and buildings (figure 9.5.a). Roads must be extracted and represented by their medial lines and buildings by their contours. These objects therefore really need to be separated and described using different vectorisation methods. One way to separate them is to analyse their parameters in the raster image or to apply, for example, the morphological operations discussed in Chapter 5. The alternative is to extract and examine object contours. Below we describe an instance of this latter approach.

The second map layer incorporating roads is the black layer (figure 9.5.b). The black layer contains two types of roads, "thin" and "thick". A single line represents thin roads while parallel lines corresponding to their sides mark thick roads. Both types must be represented via medial lines. In our system, the black layer is vectorised using a variant of the raster to vector transformation based on skeletonisation. The result is represented in the form described in Section 7.6 and provides the input to the road recovery process. We now consider in further detail the extraction of roads from these two layers.

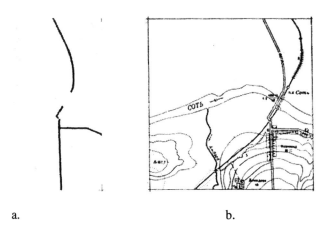

Figure 9.5. Examples of a) the road layer and b) the black layer of a topographic map.

9.2.2 Interpreting the Road Layer

We underpin our interpretation of images of the road layer with a variant of the raster to vector transformation based on contouring. The road layer interpretation algorithm may be summarised thus [200]:

1. Identify and vectorise the boundaries of all image objects.
2. Classify all vectorised contours as either elongated (roads) or not (buildings).
3. Extract medial lines from elongated objects.
4. Record results in an output database.

Step 1 comprises the extraction of object contours and their transformation into vector form via simultaneous contour approximation and the calculation of object characteristics. Relations between objects may be computed if necessary. These operations were described in some detail in Section 6.3. To distinguish elongated from compact objects, the average thickness of each connected component is first obtained. This is achieved by assuming that the connected component comprises a set of trapezoids, then the average width is given by:

$$W = 0.25*(P-sqrt(P*P-16*S)),$$

where S and P are the area and perimeter, respectively, of the connected component. Application of a simple threshold to W distinguishes road from building objects.

The algorithm that extracts medial lines from elongated objects proceeds as follows:

- Start at one end of an elongated object.
- Simultaneously trace both sides of the object, extracting the medial line and identifying branches on the way.
- Separate branches from the central region and store the location of their ends.
- When the end of the current region is reached, process all the branches identified in the same way.

This is similar in spirit to the line tracking algorithms discussed previously. The peculiarity of this algorithm, however, lies in its use of special marks to identify places at which branches should be cut from the contour.

The beginning of an elongated object is located by seeking two vectors $V_1 = (t_0,t_1)$ and $V_2 = (t_2,t_3)$ which belong to one contour and satisfy the following conditions (figure 9.6):

- angle 1 between vector (t_1,t_0) and vector(t_1,t'), where t' is the contour point following t_1, is not more than 180 degrees;
- the cosine of angle 2 between V_1 and V_2 is less than -0.8;
- at least one perpendicular may be dropped from the end of V_1 onto V_2 or vice versa;
- the length of these perpendiculars is not more than the maximum width of the object;
- all the contour points between t_1 and t_2 are inside of a square centred on point t_1 with side length W, where W is the distance between V_1 and V_2;
- point t_2 is to the right of vector V_1.

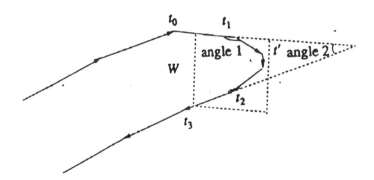

Figure 9.6. Identifying the beginning of an elongated object.

Tracing of the elongated object starts from the points t_1 and t_2 and follows two co-ordinated paths along the two opposite sides of the road, simultaneously extracting a medial line and its geometric approximation (figure 9.7). Movement from t_2 traverses the contour in a clockwise direction, while movement from t_1 proceeds anti-clockwise. Both paths are extended simultaneously; selection of the next point to move to from t_2 is influenced by the possible moves from t_1 and vice versa.

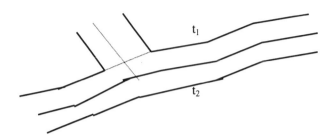

Figure 9.7. Contour tracing with medial line extraction and interpretation of branches.

During tracing, the following situations are detected and characterised:

- convergence of the traced contours onto a common point;
- divergence of the contours beyond an acceptable level;
- return to a previously marked branch.

Convergence implies that the end of the road has been reached and that its skeleton should be written to the output database. When divergence is detected, further movement is made along the two sides of the contour. Tracking continues as long as the distance between opposite points remains less than the maximum allowed road width. If the maximum width is exceeded the contour is cut; the point is marked, skeletonisation is terminated and the algorithm seeks the beginning of a new object. The third and final situation arises when the tracker revisits a point at which such a cut has already been made and marked. When this occurs the currently tracked object is itself cut and tracking is restarted from the mark. Processing continues in this way until all branches have been considered and a complete vector description produced (Chapter 7).

9.2.3 Extracting Roads and Correlated Objects from the Black Layer

Following the principles outlined in Chapter 8, we interpret the black layer of a map overlay only after having applied the processes presented above to the corresponding road layer. Input to the algorithm seeking to extract roads from the black layer are the vector-based road model, $Mv(r)$ (figure 9.8.a), obtained from the previously processed road layer and a vector description, $Mv(b)$, of the black layer (figure 9.8.b). $Mv(r)$ contains road skeletons while $Mv(b)$ describes both sides of each road. $Mv(b)$ is obtained and represented as described in Chapter 7.

Recognising Cartographic Objects

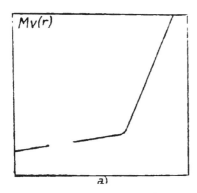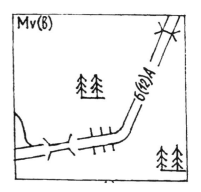

Figure 9.8. A fragment of a) a road layer and b) the corresponding black layer.

The aim of the procedure described here is to extract and identify both roads themselves and correlated roadside objects. Consider first the extraction of roads; the process is as follows [200]:

1. Establish a geometric correspondence between Mv(r) and Mv(b) by scaling, rotating and translating Mv(r) as necessary.

2. Use roads from Mv(r) to identify primitives likely to describe roads and roadside objects in Mv(b). All the black layer primitives situated in the neighbourhood of a road layer road (figure 9.9) are marked and collected together in a road list (or r-list). Primitive characteristics such as length, orientation, distances between end points, etc. are computed at the same time and added to the r-list representation.

3. Trace and combine primitives in Mv(b) to form roads, labelling adjoining primitives as likely to describe roadside objects (figure 9.10).

4. Identify gaps in the roads detected in Mv(b), which are likely to contain road signatures (identifying numbers).

5. Connect the road segments abutting each side of the gaps, interpolating to create a complete description of the road in the r-list.

6. Extract, as accurately as possible, the medial line from the road in Mv(b). This will form the basis of the final road representation. Road skeletons extracted from the road layer could be used to underpin the final representation. However, positional information obtained from the black layer is generally more accurate and is therefore used in our system.

7. Identify any primitives (i.e. vectors) appearing within road gaps that appear to arise from a road signature.

8. Recognise the road signature in the gap.

9. Create the final road representation, recording its type, characteristics and metrics.

This procedure is repeated for every road in Mv(r).

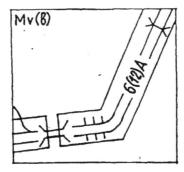

Figure 9.9 Primitives arising from roads and correlated roadside objects

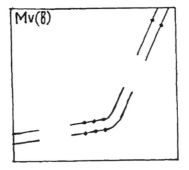

Figure 9.10 Primitives arising from roads only

On termination of the road recognition process, attention turns to those primitives stored in the r-list which describe roadside objects (figure 9.11). A priori knowledge of such objects (bridges, buildings, etc.) is used to establish relationships between accepted roads and remaining primitives. The knowledge base allows the system to reduce the quantity of objects considered and so helps to minimise recognition time. Correlated object recognition proceeds thus:

1. Primitives having a specific relationship to a recognised road are first extracted from the r-list and their representation extended to include some additional parameters. The basic relationships sought are: joining, crossing, linked-by-gap and lie-in-neighbourhood. Once a relationship has been identified, a local road situation is defined. Using the local road situation and associated primitive parameters a hypothesis is advanced regarding the possible class of the new object.

2. Primitives associated with each correlated object are clustered together using a priori knowledge of the hypothesised object class.

3. The hypothesis is confirmed (or rejected) by a structured pattern recognition process which matches the extracted primitives to a class exemplar.

4. Geometric properties of the recognised object are computed.

Figure 9.11. Primitives arising from roadside objects

After analysis of one local road situation is completed, attention turns to the next. Subsequent objects are sought alongside roads crossing either a previously considered road or a new road from the road layer. The process continues until there are no unrecognised objects in Mv(r). Figure 9.12.b shows the roads extracted from the road layer of figure 9.12.a.

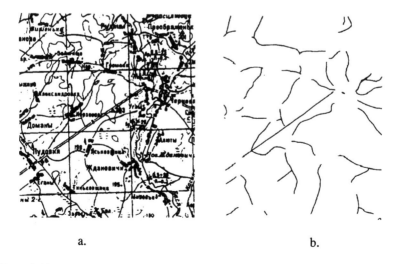

a.   b.

Figure 9.12. a) A topographic map and b) the roads extracted from the road layer of a).

9.3 Recognising Texture and Area Objects

9.3.1 Overview

The interpretation of textured objects on line drawings is naturally split into two parts. The interpretation system must both recognise the type(s) of texture on the drawing and describe the borders of textured regions.

Three primary approaches to texture segmentation are identifiable in the literature. The first seeks to produce efficient and effective segmentation algorithms by studying and attempting to mimic human perception of texture [201]. This is of considerable scientific interest, though the techniques it has produced to date are not widely used in line drawing or map interpretation. The second approach is to focus on the internal properties of textured regions, developing texture characterisation and identification tools [202,203]. The third concentrates on the analysis of texture boundaries [204]. The last approach relies primarily on statistical tools. The methods developed perform well, but tend to be computationally somewhat expensive.

Texture-based image segmentation involves grouping image data into regions with similar textural properties and then extracting their borders. Most texture segmentation algorithms perform a statistical classification after transforming the image intensity data into some feature space. Texture features are typically multidimensional, making the histogram-based thresholding methods discussed in Chapter 4 difficult to apply. More complex methods, often employing Bayesian statistics and nearest-neighbour analysis, are more likely to be applicable. Prior knowledge of the number and types of texture in the image is often required. Most algorithms consider grey level images and do not take the special properties of line drawings into account.

The identification of texture boundaries is made difficult by the large variations in the properties of the textured regions that generate those boundaries. Borders between homogeneous regions of different grey level are typically extracted using only intensity information and are based on edge detection methods. Local intensity information is not sufficient, however, to determine texture borders. When edge detectors are applied to binary textures in an effort to find borders the result is at best unpredictable. Processes that aim to identify texture borders must in some way take into account the properties of the classes of texture they separate.

9.3.2 Texture Border Definitions

Contours bounding image regions are variously defined in the literature using terms such as boundary, border and edge. Each has a particular, precise meaning. In general, the boundary passes between adjacent pixels; one of which belongs to the object region and the other to the background. The boundary is of necessity a closed curve. The border is the set of background pixels bounding the object region. Conversely the edge is the set of external object pixels, those which are adjacent to background pixels.

The limits of textured regions are harder to define. Each region comprises a set, not of pixels, but of texture elements, which in macrotexture could be of significant size. Where, then, is the line between textured object and differently textured background to be drawn? This question has no unique answer.

Consider the limits of the binary texture shown in figure 9.13. One way to specify a texture boundary is to create a polygon whose vertices are the centres (in general the centres of gravity) of the external texture elements. This is the first type of boundary shown in figure 9.13. The second is the convex hull; the shortest path enclosing all the pixels that contribute to some texture element within the region. The third type of border is obtained when region growing operations such as morphological dilation are applied to the black pixels of each texture element until the texture region becomes a single connected region of black pixels. Boundary location then becomes a standard problem in binary image processing. The fourth and final example shown in figure 9.13 is the convex hull of the type 3 border.

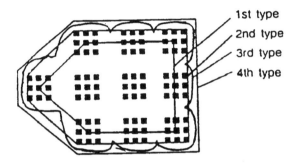

Figure 9.13. Four types of texture border.

It is clearly possible to define other classes of texture border, though most will appear somewhere between the first and fourth above. In general we impose the following requirements on any texture border reported during map interpretation [172]; a texture border must:

- be close in some well-defined sense to the external texture elements of some region;
- be of minimal length;
- be simple to compute;
- correspond reasonably well to the border extracted by a human observer;
- given descriptions of the texture patterns it separates, allow the local image texture to be reconstructed.

9.3.3 A Texture Border Extraction Algorithm

The extraction of texture borders has been approached in many ways, often drawing on previous work on the segmentation of grey level or binary images. The distance transform (Chapter 5) can be exploited by applying it to the background (white) areas of a textured map image and examining the resulting distance map. Distance maps obtained in this way contain a set of waves, some of which can act as texture borders (figure 9.14). As distance from the black regions comprising the texture element increases, contours of constant distance will begin to emerge. These contours can provide adequate boundaries for compact convex textured regions. The distance at which the boundary is considered to lie should vary with the shape and density of the texture concerned [172].

-3	-3	-3	-3	-3	-4	-4	-4	-4	-4
-2	-2	-2	-2	-2	-3	-3	-3	-3	-3
-2	-1	-1	-1	-2	-2	-2	-2	-2	-2
-2	-1		-1	-1	-1	-1	-1	-1	-2
-2	-1			-1	-1			-1	-2
-2	-1		-1	-1	-1			-1	-2
-2	-1	-1	-1	-2	-1	-1	-1	-1	-2
-2	-2	-2	-2	-2	-2	-2	-2	-2	-2
-3	-3	-3	-3	-3	-3	-3	-3	-3	-3
-4	-4	-4	-4	-4	-4	-4	-4	-4	-4

Figure 9.14. Waves on the distance map (an example is shown in bold) can form the border of a simple textured area.

A more generally applicable, albeit more complex, border extraction algorithm exploiting the distance transform can be summarised as follows:

1. Apply the distance transform to the image background, recording distance as a negative value (Chapter 5).

2. Select a threshold value -h and set all elements in the distance map with absolute less than h to zero, producing a channel of zeroes between the black pixels and the -h-wave. Figure 9.15 shows the result of applying this operation (h=-3) to the distance map shown in figure 9.14.

-3	-3	-3	-3	-3	-4	-4	-4	-4	-4
0	0	0	0	0	-3	-3	-3	-3	-3
0	0	0	0	0	0	0	0	0	0
0	0		0	0	0	0	0	0	0
0	0			0	0			0	0
0	0		0	0	0			0	0
0	0	0	0	0	0	0	0	0	0
0	0	0	0	0	0	0	0	0	0
-3	-3	-3	-3	-3	-3	-3	-3	-3	-3
-4	-4	-4	-4	-4	-4	-4	-4	-4	-4

Figure 9.15. Application of a threshold (h=-3) to the distance map of figure 9.14.

3. Compute the reverse distance transform, propagating distance values in the inward direction away from the wave and towards the texture elements.

Recognising Cartographic Objects

At the end of this process pixels with positive (i.e. black) and zero values are bounded by (-1)-waves (figure 9.16). These waves both surround and lie close to textured areas and can be used as texture borders. Figure 9.17 shows the texture borders extracted from map areas depicting a river and a sandy area.

-3	-3	-3	-3	-3	-4	-4	-4	-4	-4
-2	-2	-2	-2	-2	-3	-3	-3	-3	-3
-2	**-1**	**-1**	**-1**	-2	-2	-2	-2	-2	-2
-2	**-1**		**-1**	**-1**	**-1**	**-1**	**-1**	**-1**	-2
-2	**-1**			0	0			**-1**	-2
-2	**-1**		0	0	0			**-1**	-2
-2	**-1**	**-1**	**-1**	**-1**	**-1**	**-1**	**-1**	**-1**	-2
-2	-2	-2	-2	-2	-2	-2	-2	-2	-2
-3	-3	-3	-3	-3	-3	-3	-3	-3	-3
-4	-4	-4	-4	-4	-4	-4	-4	-4	-4

Figure 9.16. Distance map of figure 9.15 after application of the reverse distance transform. Bold pixels show the proposed texture border.

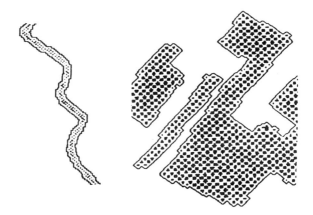

Figure 9.17. The extraction of borders from textured areas of an input map.

As roads, railway lines, buildings, etc. can cross or appear within, for example, forests and areas of sand, it is common for textured map regions to contain or otherwise interact with sizeable black components. The boundary detection algorithm above can be extended to extract black areas by applying a separate distance transform (DT_1) to figure and background pixels (DT_0). Two thresholds, h_1 for DT_1 and h_0 for DT_0 are required. The value of h_1 is chosen to be greater than $T/2$, where T is the average of the widths of the lines comprising any line and net textures and the diameters of the black blobs making up any point textures. The algorithm can then be written as follows:

- Compute the distance transform over both figure and background, assigning object pixels positive and background pixels negative distance values.

- Identify h_0- and h_1-waves.

- If the h_1-wave lies inside the h_0-wave take the wave at $h = +1$ to be the black area border; otherwise the h_0-wave is the border of a textured region.

Further texture border extraction algorithms based on the distance transform are given in [172].

Having identified texture borders, the type of texture inside each region must be identified. For simple cases this can be achieved using a knowledge base recording the characteristics of the various expected texture elements. Features such as the ratio of black to white pixels within the texture element, expected sizes of the connected components making up the texture element and/or the wave numbers of the texture border might be used. In more complex situations statistical methods may be required. An example of the extraction of area objects (forests) from a topographic map is shown in figure 9.18.

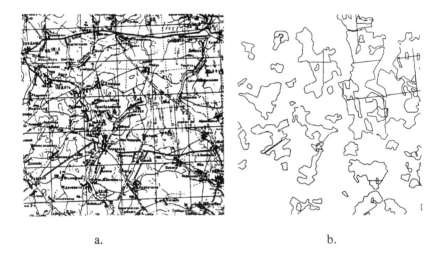

a.   b.

Figure 9.18. a) A topographic map and b) the extraction of forest from a).

9.4. Recognising Symbols

The final group of cartographic entities to be recognised comprises various symbols, characters and text. Many text recognition systems have been designed and implemented, though these are usually very specialised and take into account the specifics of the class of document being processed. A good overview of this field is given in [205]. As noted in Chapter 1, cartographic symbols are usually more complex than simple alphanumeric characters, can appear almost anywhere on the map at a variety of orientations and scales, and display numerous relationships with each other (figure 9.19). Cartographic symbol recognition is a complicated task and has not yet been solved.

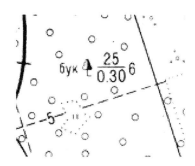

Figure 9.19. Cartographic symbols on a Russian topographic map.

A number of projects have focused on the recognition of various sets of cartographic symbols [173,175,187,206]. These all assume that the symbols have been separated from the rest of the graphics, and isolated, before recognition begins. Separation of text from graphics may be achieved on the basis of a raster image representation [207,208] or after vectorisation [209-212]. Image-based methods are the most common, though this approach can be problematic given map images. Text/graphics separation is reasonably reliable when text and graphics form well-localised, compact areas: on maps the text and symbols are much less constrained.

Cartographic symbol recognition usually focuses on symbol structure, with recognition being viewed as the search for particular patterns of symbol features. Two representations are in general use: contours and skeletons (Chapter 6). A particularly effective method based on contours is described by Elliman and Sen-Gupta [187] and has been tested on a large set of cartographic symbols. In contrast, we have developed an approach to the recognition of cartographic symbols, at certain standard orientations, based on structural parameters of skeletons [193]. This representation was adopted because, while their area properties frequently vary from one map to another, cartographic symbols generally have well-defined and consistent skeletons. Taking the standard structural pattern recognition approach, primitives and features are first identified. The features employed are:

- Object type. The object could be either simple, comprising only one segment, or complex, consisting of several segments. Simple objects may be divided into two subgroups. The first contains open objects with two end points (figure 9.20.a), the second contains closed objects (figure 9.20.b). Complex objects may also be subdivided, separating those which include internal contours (figure 9.20.c) from those which do not (figure 9.20.d). Each input object is assigned a code representing its position in this simple hierarchy.

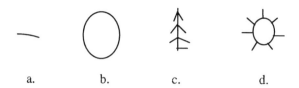

a.   b.   c.   d.

Figure 9.20. Examples of object types

- Number of internal regions (holes) in the object.

- Number of segments comprising the object and some very simple orientation statistics; at present a simple count is made of the number of segments lying at 0, 45, 90, and 135 degrees to the horizontal.

- Number of nodes and end points and their distribution within a bounding rectangle. The bounding rectangle is divided into four equal parts; the number of nodes in each and the total are recorded.

- Object length and height, i.e. the dimensions of the bounding rectangle.

These features are combined to produce a feature vector representing the input skeleton. This vector can be thought of as specifying a point in an n-dimensional feature space. The image co-ordinates of one or two points on the symbol are also computed and stored in a location vector. Two points are stored when specific symbols must be represented by the locations of two of their (variable position) features in the output database.

The feature vector is then compared with a set of reference vectors stored in a symbol library. Comparison and classification is achieved using one of a number of possible similarity measures. One method is based on the k-nearest neighbour approach. A modification called weighted k-nearest neighbour, which takes into account the relative importance of features was proposed in [213]. The k (chosen empirically) library vectors most similar to the input vector are located. Each is given a vote, whose value varies inversely with its distance from the input vector according to the voting formula:

$$\text{Vote}^L = 1 / (\text{dist}(F^L, F^I))^2$$

where F^L and F^I are feature vectors describing the library instance and the input symbol, respectively. The weighted Euclidean distance between the two vectors

$$\text{dist}(F^L, F^I) = \text{sqrt}(\, w_i \, (F^L_i - F^I_i)^2\,)$$

is computed, where F^L_i is the ith feature of the library vector, F^I_i is the ith feature of the input vector, and w_i is the ith weighting factor. This weighting factor is designed so that features with smaller variance are given a larger weight. The votes for each class C are summed and a certainty value is computed for each class C_i found among the k-nearest neighbours. This value approximates the certainty that the input vector belongs to C_1 [213].

Chapter 10
Recovering Engineering Drawing Entities from Vector Data

10.1 Design Principles and System Architecture

The machine interpretation of engineering drawings has developed quickly over the last two decades. Early work focused on vectorisation [157,168,214,215], though more sophisticated entity recognition systems soon began to emerge [14,188,216-218]. Analysis of the performance of these systems and our own experience in engineering drawing interpretation suggests some general principles that might usefully be borne in mind [13].

Design criteria for drawing interpretation systems were presented in Chapter 2, and some principles underlying the interpretation of images of maps were put forward in Chapter 8. The design principles upon which engineering drawing interpretation is or should be based do not differ greatly; engineering drawing interpretation is a specialisation of graphics recognition and quite similar to map interpretation. We will not, therefore, repeat all that has gone before, but concentrate instead on notions of particular relevance to engineering drawing interpretation. Against this background, we consider a number of guidelines that support the interpretation of images of engineering drawings.

1. Proceed from the simple to the complex: i.e. start with vectorisation and move towards the more complex CAD entities. This guideline is at the very heart of the system architecture adopted here and described in the course of the present chapter. That architecture was shown in figure 2.16, but for convenience is reproduced as figure 10.1.

It should be stressed that in engineering drawing interpretation moving from simple to complex entities really is a general guideline, and not a hard and fast rule. The system described here, for example, does not try to recognise all circular arcs at once. Some are left until axes of symmetry have been identified, simply because it is easier to extract arcs whose centres are known. The system also revisits the identification of text strings after geometric primitives have been identified. It is not uncommon for geometric primitives to pass through areas of text and disrupt the text recognition process. When lines, etc. have been identified as such, there is therefore a reasonable expectation that previously extracted character strings can be extended.

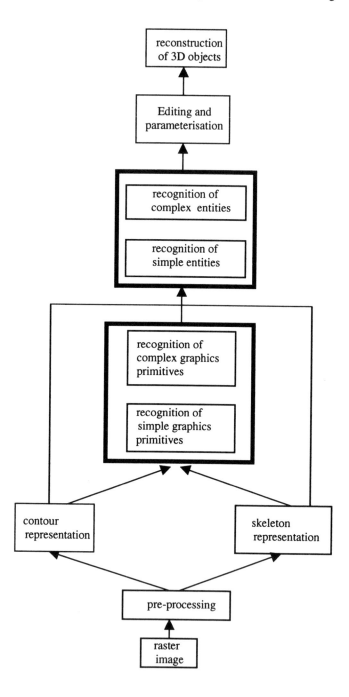

Figure 10.1. A bottom-up, sequential architecture for engineering drawing interpretation.

This being said, there remains a strong flow from the simple to the complex. Dimensioning, for example, is the most complex and therefore one of the last constructs to be considered. By postponing the search for dimensioning until we have extracted as much information from the drawing as possible we reduce the number of false positive dimension sets hypothesised (and therefore tested, at some expense).

It will also be noted that figure 10.1 appears to make a hard distinction between primitives and entities, and, within these classes, simple and complex constructs. Simple and complex primitives were discussed in Chapter 2, as were the various types of entities typically considered in map and engineering drawing interpretation systems. It should be clear from the early discussion and the examples of our own systems presented in later chapters what these terms mean to the present authors. The developers of different systems and approaches will, however, draw similar distinctions at slightly different levels of abstraction.

2. Make maximal use of any expected spatial relations between primitives and entities. Prior knowledge of likely spatial relations (e.g. relative orientation, intersections, etc.) between entities and/or their components can be used to advance hypotheses regarding the presence of further primitives or entities. Thin parallel segments, for example, when connected through nodes by a thicker line, can be enough to suggest the presence of a hatched area. Two collinear straight segments with free end points separated by a gap might form part of an axis of symmetry or a hidden contour line. Two intersecting axes of symmetry indicate a point likely to be the centre of one or more (concentric) circles. Such relations should be exploited whenever possible. Spatial relations are obviously most useful when they generate hypotheses that lead to confirmed entity recognition. Most such relations can, however, support the extraction of more than one construct; even if the first hypothesis fails, it is usually worth recording spatial relations for later use when seeking a different graphical primitive or drawing entity.

It might be argued that this guideline is also relevant to map interpretation, and this is true to some extent. The greater prominence of large-scale symbols (e.g. dimensioning) in engineering drawings, however, means that spatial relations may be predicted more reliably than in maps. Spatial relations between map entities attempt to capture general knowledge of normal conditions and practice on the ground: bridges often lie on roads, for example. Such knowledge is valuable, but heuristic; bridges can appear away from roads too. In engineering drawings spatial relations are an integral part of the communication medium and are agreed by convention. One can therefore be much more confident that they will hold in a given drawing.

3. Proceed from local to global analysis. In the early stages of entity recognition, when only a little is known of the actual content of the drawing, it is dangerous to continue pursuit of a given entity if any contradictory or conflicting evidence exists. A mistake in such circumstances can lead to unpredictable results. We therefore distinguish two broad phases of entity recognition. During the first, in a state of information hunger, only clearly interpretable, clean sections of entities are extracted. This is achieved using strictly local analyses of situations and/or features. Only when all the clean parts of the drawing have been considered and all the easily available information identified do we try to resolve more doubtful situations, attempting a more global analysis. This approach is common in engineering drawing interpretation, though methods of analysis differ (see, for example, Chapter 11).

4. Take a structural approach. As in map interpretation, we believe structural pattern recognition to be best suited to the extraction of engineering drawing primitives, entities and symbols from vector data. If anything, structural pattern recognition is more appropriate to the (more structured) domain of engineering drawings.

5. Exploit all available knowledge. This is another guideline that was introduced earlier. It is revisited here to make the point that, if anything, more global knowledge is available in engineering drawing than in map interpretation. After entity recognition, for example, we can exploit the convention that most lines are drawn in one of two orthogonal directions. The effects of any image distortion can be reduced, along with the amount of interactive post-processing required, by making near-vertical and near-horizontal lines strictly vertical and horizontal. This quite effectively corrects other graphic primitives as well, shifting their points of intersection and producing a more faithful representation of the drawing.

6. Separate knowledge and algorithms. As noted earlier, any line drawing interpretation system should ideally be organised so that knowledge of expected entities is clearly separated from the algorithms it supports. This is particularly true of engineering drawing interpretation. Most engineering drawing entities are defined by convention. While different conventions exist in different engineering disciplines, they are often broadly similar. A drawing interpretation system that could be converted easily from one discipline to another by modification of a compact and well-defined knowledge base would be a very powerful tool.

In presenting these guidelines we appreciate that it is difficult to take into account all the relevant properties of engineering drawings. In principle one could, for example, exploit axes of symmetry when determining the image location of the physical outlines of drawn objects. To attempt this, however, would be to take a step into the larger (and more complex) problem of 3D model reconstruction from semantic information. Though fascinating, this problem is more relevant to text understanding and geometrical modelling than graphics recognition.

10.2 Vectorisation and Entity Recognition Processes

Figure 10.1 provides a useful, but greatly simplified, view of the engineering drawing interpretation system described here. In particular, it implies that drawing interpretation is achieved in a single pass through the sequence of processes shown. This is not the case. While processing (almost) always moves away from the raster image and towards the recognition of high-level and 3D objects, more than one pass through the sequence is required if the system is to deal with anything like the full range of entities that can appear on an engineering drawing. Chapters 4-7 discuss the various options available when implementing a raster to vector transformation. Each drawing type, however, has features that require fine tuning of, and perhaps small modifications to be made to, the standard techniques. The same is true of entity recognition methods. Before describing in any detail the operation of individual interpretation/recognition processes used here, we overview the operation of the system by considering the various ways in which the outline diagram of figure 10.1 is traversed.

Recovering Engineering Drawing Entities

We assume that the input drawing has been scanned, binarised and represented in modified run-length form (Chapter 3). For the images considered here a simple manual global thresholding scheme was used to produce binary images. Any of the techniques outlined in Chapter 4 could, however, be employed, without modification, to the techniques described below. Noise is then reduced by application of the techniques described in Sections 5.5 and 5.6. This level of pre-processing is standard in the system.

At this stage, after noise reduction and prior to vectorisation, text elements may be detected. The separation of text from graphics is a well-known and studied problem. Most of the solutions currently proposed, however, are intended for application to images of documents (the pages of this book, for example) in which text and graphics form distinct, independent, isolated zones. The situation is very different in engineering drawing images. Here, text and graphics are mixed and it is generally impossible to segment the image into the rectangular text-only and graphics-only regions assumed to exist in other areas of document image processing [219-222].

We address this problem, via a contour-based representation, on the first pass through figure 10.1. The widths of image lines are first examined: local line width estimates are computed at each contour pixel and a line width histogram constructed. As it is usual for engineering drawings to comprise two lines of two distinct widths, the two most prominent peaks are in the histogram are identified. This allows the maximum width of thin lines and the minimum width of thick lines to be estimated. The contours of connected components are then extracted and some of their parameters (bounding rectangle, area and perimeter) are computed. On the basis of these values, potential text symbols are identified. Individual characters are then assembled to hypothesise text strings, depending upon the patterns of spacing between them [13]. A typical engineering drawing image input to the system described here is shown in figure 10.2.a. Figure 10.2.b shows the text strings extracted from it by the algorithm outlined here and described in detail in [13]. The recovery of other entities from this sample from this drawing is illustrated as the chapter progresses.

Referring to figure 10.1, text strings may be considered complex entities, individual text symbols simple entities and the contours of connected components simple graphical primitives. The extraction of complex graphical primitives is not necessary in this part of the task.

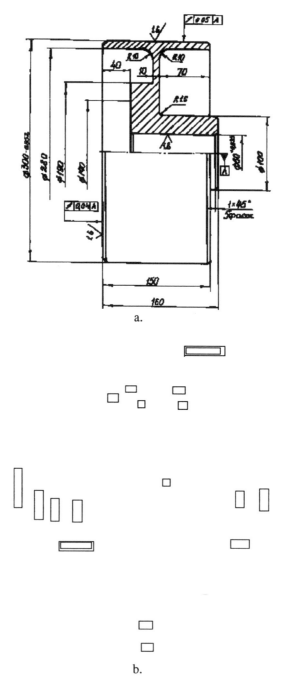

Figure 10.2. a) An image of a typical engineering drawing processed by the system described here and b) the text strings extracted from it.

Recovering Engineering Drawing Entities 191

We now consider the second pass through figure 10.1. The line width estimates gathered in the first pass are now used to parameterise further pre-processing in the form of morphological erosion. A circular structuring element is applied with radius equal to half the maximum thin line width plus one. Thin lines are deleted, leaving only arrowheads behind. These are then subtracted from the original image to produce a further image containing only thin lines (figure 10.3.b). Arrowheads are then reconstructed, on the first image, forming isolated objects that are easily recognised by the distinctive shape of their contours (figure 10.3.c). If the image is expected to contain large black areas, further, similarly parameterised, erosion is used to delete thick lines. Dilation is then applied to reconstruct larger connected components. Having separated connected components corresponding to thin lines, thick lines, arrowheads and larger regions, a third pass takes the alternative, skeleton-based path through figure 10.1.

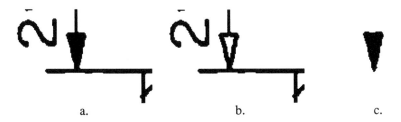

Figure 10.3.a) An engineering drawing image and the recovery of b) thin lines, c) arrowheads from a).

In the third pass, connected components expected to represent thin and thick lines are thinned and line width estimated at each skeleton pixel. Width values obtained at this stage may reasonably be expected to be more accurate and reliable than those reported during the first pass. Feature points (nodes and ends) and segments bounded by feature points are then extracted from the thinned image. Each segment is vectorised and approximated (Chapter 7). Segment properties like average width are computed and structural noise reduced by examining each segment's length and type (Chapter 7). The skeleton representation is completed when geometrical primitives are fitted to each segment. Critical points (at which local curvature changes significantly) are extracted and skeleton sections bounded by critical points are approximated by lines and arcs.

Neighbouring geometrical primitives are then examined to see if they can be merged together; it is not uncommon for critical point detection to generate the occasional false positive. Arcs and other complex graphical primitives are also fitted to connected straight segments. A simple tree search drives the process. For each simple geometrical primitive, a tree of possible paths through the skeleton (comprising only segments linked by node points) is built and analysed. The best path through the tree is identified and its components merged, where possible, to form longer or higher level graphical primitives.

The vectorisation process outlined above (Section 10.3) is also applied to the contours of other connected components, producing the skeleton and contour-based vector representations upon which entity recognition is based. The combined vector model can, however, if wished, be transformed into IGES, DXF or DXB AutoCAD formats and output as a drawing representation in its own right.

Entity recognition may now begin. Dashed lines are extracted after examination of segment shape, type and local structure. The algorithm for dashed-line extraction from maps may also be applied to engineering drawings and is described in Section 9.1. The intersections of dashed lines then denote potential circle centres. If any are found a further attempt is made to identify arcs and circles, as above. Vectors of restricted length but relatively large width are separated and an attempt made to use these to extend, and hopefully complete, text strings identified during the first pass. Crosshatched areas are also extracted (Section 10.4). Thick lines which might lie on the boundary of a hatched area are located (Section 10.1), and joined together by a process which searches for closed borders.

Dimensions are among the last entities to be extracted (Section 10.5). Initial hypotheses are provided by text strings and potential arrowheads. Knowledge of the likely structure of and spatial relations within dimensioning is then used to check these hypotheses and propose others. At this stage, any thick lines that are not borders of hatched regions are assumed to represent the physical boundaries of drawn objects. An attempt is therefore made to link such segments together, again to form closed contours (Section 10.6).

Once entity recognition is complete, geometric correction is performed. Near-vertical or near-horizontal lines become strictly vertical or horizontal, near-parallel lines become parallel etc. At the same time, points of contact between different entities are revised. The final, entity-level description is then converted to a standard CAD output format (DXF, AutoCAD or IGES) and the system terminates.

10.3 Extracting Arcs and Straight Lines

Circular arcs and straight lines are perhaps the most important primitives in engineering drawings; it is critical that they are recognised and presented as accurately as possible in any output description. Existing methods of straight line and arc recognition may be classified broadly into two groups (see Chapter 7 for an alternative and more detailed taxonomy):

1. methods which employ a conversion from the original metric space (the image) to another space in which graphical primitives may be more easily detected, e.g. [223];

2. methods operating in the original (image) space, e.g. [177].

The first set of methods relies to a large extent on the Hough transform (Section 7.4). Major drawbacks of this approach are low noise immunity and relatively low speed when implemented on standard, sequential processors. An illustrative example of group 2 methods is the perpendicular-bisector tracing algorithm [188], which is used for arc segmentation. This algorithm examines the raster image to find three points that lie approximately on a hypothesised arc. These either allow the initial arc parameters (centre, end points and width) to be refined, or contradict the assumption that an arc exists in the examined location. Other algorithms in this style are described in [224,226]. One drawback associated with members of the second group is low versatility; such methods generally lack the ability to simultaneously identify two types of primitive without incurring significant computational overheads.

The extraction of straight lines and circular arcs from engineering drawings must be achieved by a process which is both sufficiently fast in operation and reasonably insensitive to noise. As outlined in Section 10.2, the method used here comprises two stages [13]. Simple graphical primitives are first identified, with no analysis of or reference made to, any spatial relations between them. Vectors are then aligned and joined together, subject to consideration of the interrelations between them, to form long straight lines and circular arcs. This is an important step in the interpretation of engineering drawings; straight lines are expected to be mostly horizontal or vertical and arcs should be as long as possible.

The simplest form of the algorithm is to build two trees recording all possible continuations of an arbitrarily selected line segment; one in each direction. The path through this structure (i.e. through the initial segment) for which the sum of the lengths of the segments involved is maximal is then found. That path is then traced. At every step the segment at the current head of the path is compared with its predecessor and an attempt made to merge the two lines to create a single acceptable segment. This can be achieved either by merging two descriptions at the vector level or by returning to the image to seek a new vector which subsumes those extracted previously. If at least one of the contributing lines is longer than the newly created vector, the new line is labelled invalid; the attempt has failed. If the new or composite vector is longer than both of the original vectors the attempt is judged a success and the new line is included in the output model.

The above solution is simple, but computationally quite expensive. We therefore employ a less general, but faster solution. The tree of continuations is constructed as before, but only in one direction. The longest straight line in it is then identified and used as the starting point for a tree of possible extensions in the opposite direction. Line segments are sorted on length. Longer lines are usually the more reliable and it makes intuitive sense to begin with the most reliable local data (Section 10.1). Vectors are chosen in length order and attempts made to merge them with their neighbours and/or replace them with newly extracted longer primitives as outlined above. The maximum depth of the tree must be increased if re-examination of the lower-level representations suggests that a previously extracted vector should be replaced by more than one newly created vector. By allowing this to happen we avoid the ambiguities which can arise when several primitives pass through the same segment. The maximum depth of the tree is the length of the longest possible extension. Note that longer paths may be available through the raw vector data, but not included in the tree because they form extensions classed as invalid.

A similar approach is taken for the recovery of arcs. The only difference is that several circles (touching one another) may pass through any given vector. Strings of straight segments are combined to generate arc hypotheses. Each hypothesis is then checked to see if exactly two straight lines could explain it away. If an accurate two-line approximation exists, the arc is considered invalid; otherwise it is accepted. Figure 10.4 shows the arcs extracted from a vector representation of a section of engineering drawing. When all of the arcs and straight lines are extracted, the approximation of the remaining broken lines is performed as described in Section 9.1.

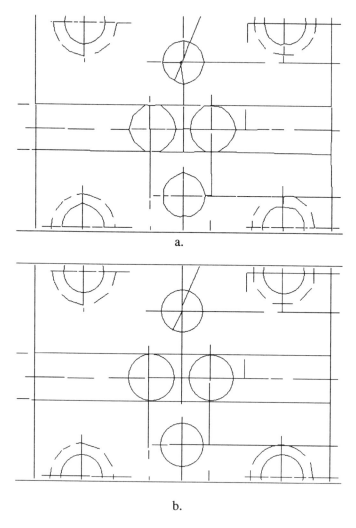

Figure 10.4 .a) A vector description of an engineering drawing and b) arc extraction from a).

Geometric correction of a drawing representation to enforce known high-level properties should really be left until all drawing constructs have been identified (Section 10.1) and entity type can be taken into account. Hatched lines should, for example, be constrained to lie at 45 degrees to the horizontal. The system cannot know which lines should be subject to which constraints until entity recognition is complete. Some correction can, however, be performed at the vector level. Specifically:

Recovering Engineering Drawing Entities

- straight lines that are nearly horizontal or vertical are made exactly horizontal or vertical;
- any points of intersection between lines altered in this way are updated;
- the centres of concentric circles are set to a common point.

This is done if and only if the vector description is all that is required in the current application. After these processes are complete we obtain an output model which can be converted to any CAD system format. Figure 10.5 shows an example of the interpretation of the engineering drawing shown in figure 10.2.a following this level of analysis.

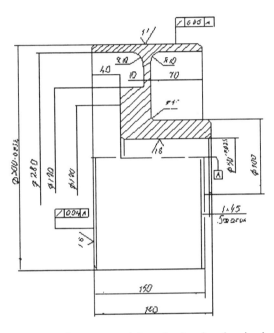

Figure 10.5. Arcs and straight lines extracted from the drawing showing in figure 10.2.

10.4 Recognising Crosshatched Areas

Crosshatched areas are a common line drawing component representing, among other features, sections on engineering drawings and buildings on maps. Though conventions obviously vary in their details, crosshatching is always a variant on the standard theme of parallel lines somehow gathered together to form closed regions.

An interesting approach to crosshatch recognition was developed for the French systems CIPLAN, REDRAW and CELESTIN [15,16,227]. Elongated connected components with common width and orientation are first extracted from the raster image. These provide a low-level image model. The image is then vectorised and crosshatched areas formed using information from both the vector database and the image model. The methods proposed rely heavily on a priori knowledge. All three systems adopt the same basic approach, but differ in their high-level processing.

In [20], crosshatched areas are recovered from raster images of land register maps. An image graph describing connected components and their interrelations is formed and used to guide the search of a closed polygonal area of the image. Any nodes and edges considered to form part of crosshatching lines are removed from the image graph. If the search for hatching fails, a human operator is asked to provide help.

We have developed an algorithm [13] which recognises crosshatched regions in engineering drawings. This technique takes as its input a vector database and operates by tracing around potential boundaries of crosshatched areas, identifying both hatching and boundary lines as it goes. A priori knowledge of document type and general rules describing the construction of hatched areas are used to guide the search. Working with vectors rather than connected components means that dependence on constant line width is reduced to a minimum. Also, because the algorithm is a part of a larger system, it can utilise information provided by other processes. In particular, information is available regarding the type of line being considered (i.e. straight, circular, dashed).

As a rule, hatched areas are distinguished by the relative widths of hatching lines (which are fine) and boundary lines (which are heavy). In real images, all line widths are subject to noticeable variations. Moreover, the original drawing may include boundaries whose widths vary widely; local sections, for example, are often drawn very bold. In our algorithm, hatched areas are recognised in two stages, proceeding from the simple to the complex (Section 10.1). In the first stage, a local analysis of relations between vectors is used to decide whether or not a given segment might belong to an area boundary or constitute a hatching line. Hypotheses are generated and tested as an attempt is made to assemble the segments into putative hatched regions satisfying rather stringent requirements. This assembly is performed while tracing likely region boundaries. In the second stage hatched areas created during stage 1, which are often only partial, are combined in an attempt to create more complete region descriptions. The search region is expanded and the criteria upon which decisions are made are relaxed, the additional contextual information allowing the system to be less rigid in its definition of what constitutes a hatched area.

The start of a hatched region is hypothesised when two segments (L) are identified which satisfy a template description of a hatched line. The two lines are required to be connected by a third segment or group of segments (B), such that a set of template relations (E) are satisfied. The template description of a simple line comprises its width, orientation with respect to the co-ordinate axes and the type of graphical primitive to which the line contributes. The hatched area template relations are as follows:

- the two hatched line segments L must be parallel;
- each segment L must be adjacent to a boundary segment B;
- the segments L must both be located on one side of B;
- it must be possible to drop a perpendicular from one segment L to the other;
- the spacing between the segments L must be within allowable limits.

The hatched line template was formulated to be in compliance with drawing convention, taking into account the empirical study of a large number of engineering drawing images.

A line pattern satisfying E(LBL) is only tentatively regarded as the beginning of a hatched area (figure 10.6). To test the hypothesis, an attempt is made to extend the pattern in both directions. An ordered pair LB or BL is chosen, the tracing direction is determined, and all vectors connected to the starting combination and lying in the search area are identified. If more than one ordered pair of type BL is connected to the starting pattern, the pair considered to provide the most likely extension to the initial hypothesis is the one which:

- points in the same direction as the area being traced;
- comprises segments whose width is consistent with the initial hypothesis;
- incorporates the boundary segment B which forms the smallest angle with the previous boundary section;
- introduces the smallest deviation in spacing between the segments L and leaves that deviation lying within the limits specified by the template.

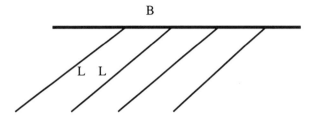

Figure 10.6. Initialising crosshatched area recognition.

A crosshatched area hypothesis is accepted if the starting combination is extended to at least the form E(LBLBL); i.e. if at least three suitably arranged hatching lines can be identified. Assembly is discontinued if:

- no further suitable segments L can be located;
- a significant change occurs in the width of the segments B;
- the chain moves outside a specified search area determined by the parameters of the developing hatched area;
- a complex branching takes place in B;
- boundary tracing locates segments which are already part of the developing region.

The data structure describing a developing cross hatched region records average orientation and average spacing between the hatching lines. Moreover, it preserves connections to the original vector model.

The second stage of crosshatching recognition takes as its input the partial hatched areas assembled in the first stage and any vectors in their local neighbourhood that have yet to be interpreted. Each hatched region is considered in turn and a search tree constructed, starting from the first (and last) segment. The tree comprises all segments connected to the outermost B and L segments and falling within a template-specific search area. All vectors running parallel to the starting crosshatch line are considered, regardless of width, as candidate L segments. All segments connected to these L candidates are considered candidate boundary (B) lines. Note that a given segment may become both a B and an L candidate. The requirements imposed on candidates are less stringent than those used previously. As the tree develops, each additional segment is characterised by several parameters. Key measures are similarity to the most recent L, deviation (if any) from the original tracing direction, width and orientation relative to parent. For a segment considered similar to L, key parameters are the nature of its relation with its parent (intersection or adjacency), its position relative to the nearest hatching line and its area characteristics (aspect ratio) if the segment belongs to a previously assembled hatched area.

As the tree is constructed, candidate hatching lines are identified. Each likely path through the tree is analysed to see if it can be extended further. Paths are terminated if they continue beyond the allowed search area or if a previously assembled section of hatching is encountered. When all paths have been followed to termination, all the available combinations of the form BL are analysed and the most likely continuation of the current hatched area chosen. Any of three decisions can be made:

- to extend the hatched area;
- to merge hatched areas together;
- to close a hatched boundary.

When a region is extended, the data structure is assembled in the usual manner. When regions are merged, two data structures must be combined. This is a simple operation; only pointers to original segments and average values of line parameters are recorded. When a hatched boundary is closed, the hatching line must be linked to the appropriate boundary segments. When, as is often the case, a hatched area has internal hollows, internal and external boundaries must also be related.

The result of applying the processes described above to the image shown in figure 10.2. is presented in figure 10.7. The algorithm has been tuned to engineering drawings and consequently produces good results on this type of image. It should, however, be noted that the quality of the interpretation naturally depends upon the quality of an original drawing. If the input drawing suffers defects beyond those implicitly expected by the system, interactive post-processing or other operator support (Chapter 8) will be required. As the current system produces standard CAD format files, interactive editing could be done within a CAD environment.

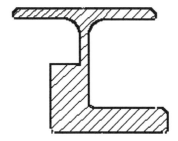

Figure 10.7. Crosshatching extracted from the image of figure 10.2.

10.5 Recognising Dimensions

Dimensions are a very important component of engineering line drawings; they provide high-level information that can influence our interpretation of large areas of the document. They provide information that can be used to check the validity of a drawing and/or to correct any errors in its interpretation. Dimensions can appear in many forms, though each generally follows a reasonably strict standard. This means that the knowledge required to construct a dimension recognition system is in most cases quite easily available. These factors have led a number of researchers to address the dimension recognition problem.

The size, shape, orientation and style of dimensioning can vary quite dramatically across a drawing. Most types of dimensioning, however, incorporate arrowheads. The first step in the recognition of dimensions is often, therefore, to locate and recognise these arrowheads [228]. Arrowheads themselves may differ in shape, size, orientation and type. They are frequently obscured by image distortion, noise and/or interfering linework. The analysis presented in [188], however, suggests that the arrowheads to be found in any given mechanical engineering drawing are usually all the same shape (triangular, rectangular or circular) and type (solid, hollow, stroke, wedge, half-filled or anchor). The self-supervised arrowhead recognition (SAR) algorithm proposed in [188] takes advantage of this uniformity by breaking recognition into two phases: parameter learning and comprehensive search. The shape, type and size of arrowhead expected are determined during parameter learning. Recognition of the entire population is attempted, using these values, during a comprehensive search.

We have developed an algorithm, outlined in Section 10.2, to extract arrowheads from raster images, that takes into account the width (in pixels) of the lines comprising the drawing [13]. A raster scan approach is adopted. Line width is first approximated in order to determine the width of stripe required (though this step may be omitted and a default stripe width used). Then, using a stripe of the appropriate size, a line width histogram is computed and local maxima corresponding to thin and thick lines identified. Morphological erosion deletes thin lines and exposes arrowheads.

Dimensions have been described using various forms of grammar and a syntactical approach has been taken to their recognition [229-234]. A dimension set is represented by a concept web whose nodes alternately represent components and the spatial relations between them. Lines are assigned attributes that are recorded in an associated line descriptor. A deterministic finite automaton is used to obtain a dimension set profile that specifies the completeness, regularity, symmetry, and type of the dimension.

A plex-grammar formalism has also been used to define a dimension model [235-237]. Component primitives are first extracted by standard low-level operations. Assembly operators are then used to combine graphic primitives. The dimension grammar comprises a set of rules describing several forms of dimensioning. A set of rewrite rules (or "productions") specifies and supports recognition of the various possible sub-shapes comprising the dimension. Longitudinal, angular and circular dimensions are identified in this way.

We have developed a dimension recovery algorithm that relies upon a priori knowledge of dimensioning stored in a knowledge base. The knowledge base has as its foundation the formal language proposed in [238] and used to describe the allowable dimension sets of an engineering drawing and a multi-level scheme for their representation. At the top level of this description is the dimension itself; text, physical outline and extension line are immediately below. These may also be subdivided; text can be split into individual characters defining, in some cases, the type of dimension, the magnitude of the dimension and its tolerances. Shape lines are decomposed into a dimension line, arrowheads and tails (extensions to the dimension line). Elements at each level are assigned descriptors, which contain descriptions of the elements in the adopted language. In the simplest cases only metric and topological features are recorded. In more complex situations relationships with other elements at lower levels of the representation must also be stored. Each component is also accompanied by a set of possible relationships to other components at the same or some higher level in the hierarchy, or elsewhere in the drawing. For example, an extension line may also be an axis of symmetry. This possibility, stated as a possible relation of the extension line, can aid interpretation in general. This type of knowledge allows all the information extracted from the drawing to be exploited at later stages in the interpretation, speeding the process and making it more reliable.

In addition to the template descriptions of dimension components, formalised as descriptors and relations, the knowledge base also includes some control knowledge. This is used to supervise and direct the recognition process. In the proposed approach, recognition of dimensions comprises a search for an initial dimension component (i.e. an initial hypothesis to be advanced) followed by parsing in compliance with the formal language and control knowledge. Exploiting information gathered by previous processes may ease selection of an initial dimension component. This information includes (figure 10.8):

- potential dimension text, identified when text and graphics are separated (figure 10.8.a);
- small components that suggest the presence of arrowheads, created when linear features are separated from large black areas (figure 10.8.b);
- lines of symmetry, which may also be extension lines, detected along with straight lines and arcs (figure 10.8.c).

In the worst case, a hypothesis may arise from any line segment, provided that it matches at least one template description held in the knowledge base.

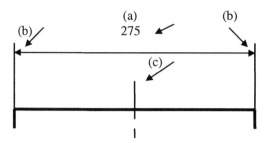

Figure 10.8. Auxiliary information used to initiate dimension recognition. See text for details

During parsing, hypotheses are advanced consecutively. Each is tested for acceptance or rejection. The formulation of hypotheses relies on knowledge of the likely relations between dimension components. A list of these, together with the pairs of dimension components between which they can arise, is given in Table 10.1. Hypotheses are accepted or rejected following reference to the template descriptions of dimension components held in the knowledge base. In the simplest case the parameters of every element of the vector model are compared with those in the corresponding template. In more complicated situations, the construction of certain topological structures may be required. For example, the template description of an extension line contains only two parameters, thickness and curvature. The description of a shape line, on the other hand, includes several grammatical structures expressed in the formal language. If rectangular blocks are used to denote dimension components and straight lines are used to indicate the relations between them, the simplest template description generates the topological structure shown in Figure 10.9.

Relation	Pairs of dimension components
Lies above	text - shape line
	text - dimension line
	text - tail
Lies on the axis	arrowhead - shape line
	arrowhead - dimension line
	arrowhead - tail
Adjoins (intersects)	extension line - shape line at right angle
	extension line - dimension line
	extension line - tail
	extension line - arrowhead
Adjoins and prolongs	dimension line – furthest point on tail

Table 10.1. Relations between pairs of dimension elements

Figure 10.9. The simplest template description of a dimension.

Acceptance of a hypothesis implies that parsing should be continued at the next level down. Parsing then begins at the newly recognised component and continues until all the dimension's components are identified. If a hypothesis is rejected, control returns to the last component on the previous parsing level. An attempt is then made to formulate a new hypothesis. If, on returning to the initial element, it is impossible to form any new hypotheses, recognition is terminated. Parsing is directed by a set of IF-THEN rules forming the system's control knowledge. These refer to the parameters of the vector elements being analysed, template values, test results, etc.

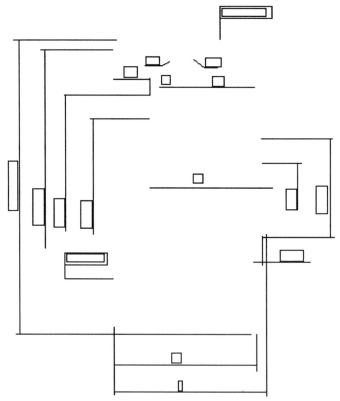

Figure 10.10. Dimensioning extracted from the engineering drawing shown in figure 10.2.

Figure 10.10 shows the dimensioning extracted in this way from the engineering drawing shown in figure 10.2. Experiments suggest that the algorithm is capable of extracting dimensioning from such images with a high level of reliability. This is because it exploits both a priori knowledge and the results of other entity recognition subsystems.

10.6 Detecting Blocks

Among the tasks to be performed by systems that seek to interpret images of technical documents is the detection of the minimal closed polygons known as blocks. In engineering drawings, blocks are usually executed in heavy lines and carry a variety of messages according to circumstances. They might represent the boundaries of visible planes, sections, or sectional views in mechanical engineering drawings, or conventional symbols in circuit diagrams. Blocks are usually detected in the closing stages of line drawing interpretation. This task has been examined in [15,239].

The classic algorithm for the detection of closed loops in graphs mentioned in [240] can be used to identify blocks and is effective, in our view, given original line drawings of good quality. In particular, the line segments forming blocks should exhibit uniform (or very near uniform) width. In real drawings, segment width may vary. If a given line is to be interpreted correctly, additional information is required. This may be obtained either during earlier processing stages or while blocks are being extracted. The algorithm we propose detects blocks by considering many properties of the available segments, generating several alternative extensions and choosing the best. Blocks are assembled as follows:

- An initial segment is chosen.
- The direction in which the block is to be traced is determined.
- A chain of segments is traced in the chosen direction.
- The resulting block is entered into the output model.

We now consider the steps listed above in a little more detail.

The initial segment must be one whose width is such that it can reasonably be called heavy, though of course its length must be greater than its width. It must extend in both directions, through nodes, and these extensions must also be heavy lines. Finally, it must not be of a type which excludes it from the block; that is, it must not be a hatching line, dimension line, etc.

Next, the direction in which a closed polygon is to be traced must be determined. The tracking direction can be either clockwise or counter-clockwise depending upon the initial segment. Direction checking and determination is done every time a segment is added to the chain.

When the path branches, offering several equivalent extensions, the segment is chosen which forces the polygon to be traced clockwise. An example is given in figure 10.11.a. If more than one such segment is available, the algorithm chooses the one which makes the smallest angle with the current segment.

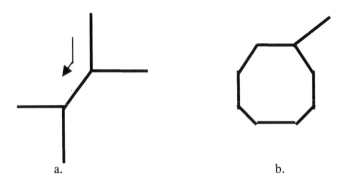

Figure 10.11. Situations which may arise during block tracing, see text.

Once the tracing direction is defined, assembly of a closed polygon begins. The path travelled during tracing and all of its adjoining segments are recorded. At each stage, segments meeting the most recently identified segment of the developing polygon at a node point are considered candidates for extension. A suitable extension is selected on the basis of:

- the width of the segments concerned;
- the type of node involved;
- whether or not a given segment is already connected to the block;
- whether or not a given segment is of a type which contradicts the hypothesis being tested;
- the angle between the candidate segment and the current most recently identified segment in the path.

This local analysis has one of two outcomes: there may either be one or several candidates for extension.

When there are multiple possible extensions a search must be undertaken. A tree of likely extensions is constructed to a specified depth. The aggregate features of the paths are determined and the alternative that is closest to the path being traced is chosen. If two possible extensions have equivalent aggregate features, preference is given to the path in which the first segment of the extension joins the last segment of the current path at the smaller angle. A block is considered complete when an extension coincides with the initial segment of the current block. If the final segment does not line up with the initial one, the path may be divided into a closed and an open part. The closed part may form a valid block (see figure 10.11.b). The open part is considered invalid and is marked as such to ensure that it is not included in any subsequent tracing. The path is also marked if it ends in an impasse; i.e. if it has no extension or includes segments whose object descriptors run counter to the hypothesis being tested.

When a block has been successfully assembled it is written into the output model and its component segments analysed. The result of this analysis is a semantic characterisation of the block. It may be marked as empty, hatched (if it contains thin parallel segments), threaded or be associated with axes of symmetry. The result of applying these processes to the image of figure 10.2 is shown in figure 10.12.

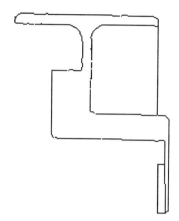

Figure 10.12. Contour blocks extracted from the drawing shown in figure 10.2.

Figure 10.13 shows the result of applying the processes described above to a more complex engineering drawing. The drawing image is shown in figure 10.13.a and its vector representation in figure 10.13.b. Crosshatched areas and blocks extracted from the image are shown in figure 10.13.c, while figure 10.13.d shows dimensions and text.

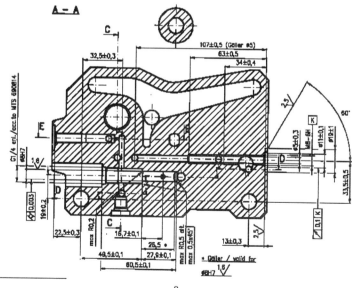

a.

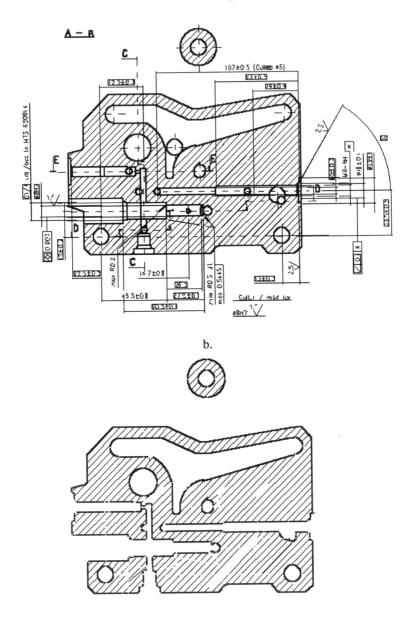

b.

c.

Recovering Engineering Drawing Entities

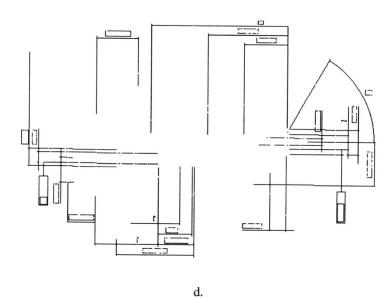

d.

Figure 10.13. a) A more complex engineering drawing, b) in vector form, c) crosshatched areas and blocks, d) dimensions and text.

Chapter 11
Knowledge-Directed Interpretation of Engineering Drawings

11.1 An Image Understanding Approach

It was noted in Chapter 1 that the interpretation of images of line drawings might usefully be thought of as an example of a knowledge-based image understanding problem. Knowledge-based vision or image interpretation systems seek to apply a priori knowledge to segment the input image(s) into regions corresponding to objects or constructs of interest in the domain at hand, providing as rich a description of those objects and/or constructs as the available knowledge allows. Despite the obvious relevance of work in this area, review of the relevant literature shows a clear distinction between attempts to improve the commercial state of the art in line drawing interpretation and more general research into image understanding systems.

While several attempts have been made to address drawing interpretation problems within knowledge-based image understanding frameworks (see Section 11.10 below) most of those interested in improving the performance of line drawing interpretation systems have prioritised the conversion of large complex images to comparatively low-level vector representations. It is often found that only when lower-level techniques have been developed as far as possible does higher-level interpretation become of interest. Workers in image understanding, in contrast, have raised interpretation to a more adequate level, by addressing higher level issues from the outset, but generally have considered only quite simple domains and images. It is not at all clear that any present image understanding system will scale up to the challenge posed by engineering drawing or map image interpretation.

In what follows we describe and discuss the ANON system [14, 241-246]. ANON arose from the belief that if we are ultimately to raise the level of interpretation in commercial line drawing systems, methods and ideas from the image understanding literature must be exploited. The assumption behind the ANON project was that the combination of these methods and ideas with a critique of existing drawing interpretation techniques would provide the basis for construction of a viable (automatic) drawing interpretation system. ANON was to be concerned only with the interpretation of images of mechanical piece-part drawings, typical input to the system is shown in figure 11.1, though it was hoped that the architecture and infrastructure developed to deal with these images would prove applicable to other drawing types.

210   Machine Interpretation of Line Drawing Images

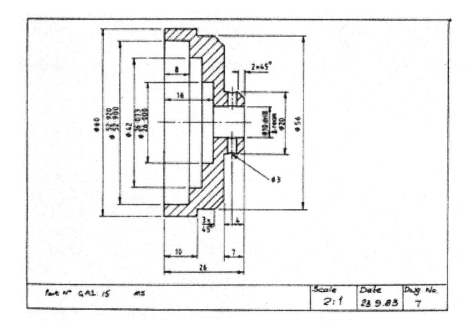

Figure 11.1. A small engineering drawing.

From an image understanding point of view the sequential architectures upon which so many line drawing interpretation systems are based is somewhat simplistic. As seen in Chapter 10, close examination of systems nominally based upon the sequential model (e.g. figure 10.1) shows that they often deviate from it. This may be to avoid some domain-specific problem that may arise when a system would otherwise be forced by its architecture to make an important decision too early. Alternatively it might be to ease development by taking maximum advantage of the knowledge and information available within the system.

The central tenet of the ANON project was that many of the limitations of current drawing interpretation systems are the result of their reliance on inappropriate architectures. In particular, a strictly bottom-up style of processing is neither powerful nor flexible enough for the task at hand [246]. When vectorisation is achieved by thinning followed by tracking of the skeletonised image, for example, variations in line width often generate short spurs that must be eliminated (Chapter 7). Careless removal of short-line data, however, can result in the loss of vital information. It is important, for example, not to discard spurs generated by thinning the arrowheads on a dimension line. This type of mistake can only be avoided by somehow incorporating into the processing knowledge of the structure of mechanical drawings.

The premise underlying the development of ANON was that the <u>entire</u> conversion process should be integrated within a suitable knowledge-based environment. Consider the intersection of broken lines: these may form L or + shapes that conform closely to those found in text strings. A detailed and high-level interpretation of the surrounding drawing is required to resolve this ambiguity. In

Knowledge-directed Interpretation

such cases specialised image search procedures should be applied under the control of higher processes to extract the significant parts of the structure.

Image understanding methods and systems can provide many valuable insights into these problems. The knowledge of engineering drawings required to exploit these methods and systems is also available; there is a rich source of knowledge about drawing constructs in standards and conventions that we can draw on for a model-based approach [247]. Moreover, the CAD structures we are seeking to generate embody composition and specialisation hierarchies (Section 11.2) that are naturally represented by schema-based image understanding systems [248-254]. The importance of high-level control of image processing has also been recognised in two classic systems: GOLDIE [254] and SIGMA [253]. There are, however, particular aspects of the drawing interpretation problem that make us selective amongst this wealth of experience:

1. The need to handle large and complex images militates against methods that maintain all possible interpretations as they proceed (e.g. SPAM [251], SIGMA), and favours those that restrict the number of alternative hypotheses that are considered (e.g. MAPSEE [248]). It is simply not practical to consider all possible interpretations of even a small area of an engineering drawing (or map).

2. Accurate interpretation of technical drawings requires the strong and varied geometrical relations between drawing components to be coupled to the image search. That is, knowledge of the likely structure of the input drawing must be used to control which areas of the image are examined, and in what way they are examined. This degree of integration remains to be achieved in even the most advanced image understanding systems [255].

3. A final, and overriding, requirement is for a system engineered to be extendable, so that the endless variety of drawing constructs can be progressively brought within its scope without the code, or its space and time demands, becoming unmanageable. This is the antithesis of SIGMA, for example, in which the structure of the system is more complex than the image model it handles, and where development implies the co-ordination of seven distinct managers, mechanisms, schedulers, generators or experts [255].

Requirement 1 above encourages us to consider architectures which examine one possibility (or only a small number) at any given time. Pursuit of a particular hypothesis will naturally require the input image (and any other available representations) to be searched for supporting and/or conflicting evidence. Requirement 2 places constraints upon this search process: it must be knowledge directed, varying with the available data and current hypothesis. The determination of search strategy in general image understanding remains problematic. Line drawings, however, are communications meant to be read. This allows insights gained from studies of human cognition to be exploited. In particular, it means that line drawing interpretation is particularly susceptible to a regular accumulation of cues (primitives) in a so-called "cycle of perception" approach (see [256] and Section 11.4). As has been noted in Chapters 8 and 10, the separation of domain knowledge from the algorithms that exploit it is a useful guiding principle in line drawing interpretation. Requirement 3 above both reinforces this guideline and extends it; suggesting that the a priori knowledge held by the system should be represented in an easily extendible form.

In the remainder of this chapter we consider the design and operation of ANON. The system is based on the combination of schemata (sometimes called objects or frames) describing prototypical drawing constructs with a library of low-level image analysis routines and a set of explicit grammatical control rules applied by an LR(1) parser. These components allow ANON to integrate bottom-up and top-down processing strategies within a single, flexible framework. ANON's a priori knowledge of drawing entities is first considered in Section 11.2. Section 11.3 discusses the system's approach to image analysis. The combination of these two key components is addressed in Section 11.4, which presents the system architecture. Control of ANON is then examined in Section 11.5 before a detailed example of entity extraction is provided in Section 11.6. The combination of top-down and bottom up processing is discussed in Section 11.7 and the system's entity extraction performance examined in Section 11.8. In the later part of the ANON project the system was extended to include some scene formation capability. This is presented in Section 11.9, along with further examples of the interpretation of mechanical drawings. Finally, ANON is compared with selected image understanding systems in Section 11.10 and conclusions are drawn in Section 11.11.

11.2 Drawing Entities as Schemata

At the heart of ANON is a set of data structures describing prototypical drawing constructs. We follow Hanson and Riseman [257] in using the term schema, with its Greek plural schemata, when referring to these objects. The alternative frame [248] causes some confusion when used in an image analysis context, conflicting as it does with the term commonly used in TV and film. It should be stressed, however, that like Hanson and Riseman, our schemata are directly influenced by Minsky's frames. Each comprises a set of variables (termed "slots" in [248]) which record key parameters, properties and abilities of the corresponding real world (in our case drawing) entity. ANON contains schema classes corresponding to solid, dashed and chained lines, solid and dashed curves, crosshatching, physical outlines, text (blocks, letters and words), witness and leader lines and certain forms of dimensioning. Note that while text is identified, no attempt is made to recognise characters. ANON merely locates text and provides as much useful information as possible to an external character recognition system [258].

Each drawing entity located by ANON is represented by an instance of a particular schema class. Every schema contains a geometrical description of the construct it represents, a set of state variables noting the current condition of that representation and a number of procedures and functions written in the C language (later implementations of ANON employ C++). The latter may usefully be organised into two halves. The first handles what are effectively administrative tasks; creating an instance of a given class, accessing and adding components, modifying state

Knowledge-directed Interpretation   213

variables, etc. The second forms the interface to the system's low-level image analysis operations.

Schemata are not isolated, but form the nodes of a network in which arcs correspond to sub-part and subclass relations. The inclusion of sub-part links means that each schema effectively maintains a structural, as well as geometric, description of the entity it represents. Figure 11.2 shows the part/sub-part links that may arise. Subclass links allow schemata to inherit the properties of related entities, in object-oriented fashion. The system's class/subclass relationships are given in figure 11.3.

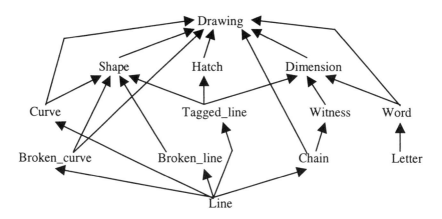

Figure 11.2. Possible sub-part relations between schemata. ANON's schemata may access functions and data structures associated with their sub-parts via these connections. Sub-parts, however, cannot access compound schemata. Note also that the above shows the set of possible part/sub-part links. The particular subset of these formed during interpretation of a given drawing depends on local context.

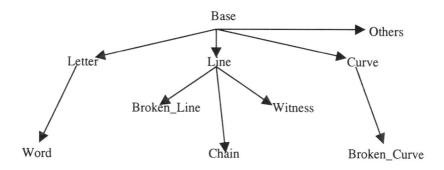

Figure 11.3. Subclass relations between schemata. These provide subclasses with a means of access to their superclass.

Consider an example. For present purposes, a chained line is an evenly spaced set of collinear line segments of constant width and contrast. These may be divided into two sets, long and short, appearing alternately along the length of the chain. The members of each set are assumed to be of constant length. Schemata describing instances of chained lines contain a geometrical description of the entity comprising its end-points and direction and the mean width and depth (contrast) of its component segments. This data structure is inherited from the schema's superclass, the line (figure 11.3). Additional summary information, particular to chains, includes mean lengths of both long and short lines and the spaces between them. A further, structural representation is provided by an ordered set of sub-part links to schemata describing the chained line's component segments.

Chained line schemata also comprise a single state variable noting whether or not the construct is likely to be ending. This is used to switch between a confident pursuit of the next dash, which tends to ignore distracting linework, and a more cautious exploration of the image that will note and report such distractions. The variable is set, for example, when segments are discovered which, while being positioned and oriented in such a way as to be considered part of the developing line, deviate significantly from the expected length and/or spacing. Such distortions are common when a chain is drawn up to a predefined point.

11.3 Image Analysis Facilities

ANON's schemata provide a common interface to a small but extendable library of image analysis routines. This collection of functions and procedures may be subdivided into those that search the image for appropriately placed ink marks and others which use such marks as seed points for the development, via direct vectorisation methods, of low-level descriptive primitives. Seeds are located using combinations of circular and linear search patterns. These composite patterns are designed to fall upon the part of the image in which the schema predicts its own continuation to lie. Examples of these patterns are given in Section 11.6 below. A schema-determined threshold is then applied to the track of pixels in the pattern and dark spots noted. These are used to initialise the tracking of straight lines (via Joseph's direct vectorisation algorithm, Chapter 7), circles [259] or area outlines [245]. In all these procedures the grey level values are compared to the grey level corresponding to the white paper background and its noise level, to give a known statistical basis for decision making at the pixel level. This normalisation is achieved via Dunn and Joseph's binarisation algorithm [64], presented and discussed in Chapter 4.

It would of course be possible to apply a knowledge-directed analysis to a pre-processed (e.g. binarised and/or thinned) image, but this would impede the deployment of a full context sensitive segmentation of the image. It would introduce a barrier to the processing sequence below which no top-down processes could be applied. It would also have an effect on the system's efficiency; in top-down analysis predictive knowledge can be used to shorten processing times by avoiding exhaustive operations such as skeletonisation.

Each schema contains functions that interface to the search routines and others that invoke appropriate tracking; both the nature and details of ANON's low-level processing depend on the class, content and state of the controlling schema. The chained line schema, for example, may perform a highly directed search for a predicted line segment involving several linear search patterns and multiple calls to line tracking. The parameters passed to these routines vary with expected line width, length, contrast, etc. (cf. [260]). If, however, the schema's state variable suggests that it might be nearing completion, a more cautious approach is taken. Circular searches (again followed by line tracking) are employed in an attempt to form a more general impression of local context (Section 11.6 below).

Note that all of ANON's image analysis is carried out under the control of some given schema, in the context of a particular hypothesis regarding the local content of the drawing. The image analysis library therefore provides an extendable toolbox of procedures whose application varies according to context. Some higher engine is needed, however, to manage the schemata; to decide what class is appropriate and what its content and state should be.

11.4 The ANON Architecture

ANON's control structure is modelled on the human cycle of perception [261]. The basis of the approach is a continuous loop (figure 11.4) in which a constantly changing world model (or schema) directs perceptual exploration, determines how its findings are to be interpreted and is modified as a result.

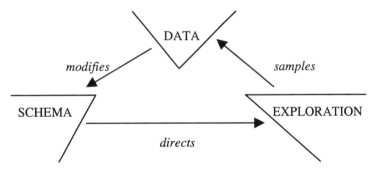

Figure 11.4. Neisser's [261] cycle of perception.

Figure 11.5 depicts the perceptual cycle as it is implemented within ANON. Each schema, as well as interfacing to low-level search and tracking routines, contains functions that examine the result of these operations and report on their findings. The controlling, or current, schema therefore both directs image analysis and interprets the result. A high-level control system is then informed of this interpretation, which takes the form of a symbolic label or token, and responds by modifying the current schema. Modification may mean updating a state variable, adding new sub-parts, or replacing the schema with a new one representing a different type of construct. Schemata representing acceptable constructs are stored in the top-level drawing schema (figure 11.2) as they are identified. All such tasks are performed by administrative methods associated with the appropriate schemata.

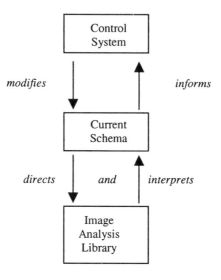

Figure 11.5. ANON's cycle of perception.

In Neisser's model of human perception the cycle continues throughout life. ANON, however, needs some procedure for initiating the system given a fresh drawing and terminating it when all the relevant constructs have been found. A bookkeeping module has therefore been designed which divides the image into 9 x 12 subareas and associates with each one an estimate of the background's grey level and noise characteristics. The bookkeeping module is initialised using Dunn and Joseph's thresholding scheme described in Chapter 4. From these initial values a target number of significantly black pixels in each area is estimated. As each new schema is stored the number of black pixels for which it accounts is calculated and subtracted from the appropriate target figure(s). The drawing schema, which is current on start-up, directs attention towards areas with large numbers of unaccounted-for pixels and terminates ANON when no significantly black areas remain unexamined.

An interesting feature of ANON's three-layer structure is its separation of spatial and symbolic processing. The system's spatial focus of attention, i.e. where it looks in the image and with what expectations of contrast, line width, etc. depends entirely upon the current schema. ANON's symbolic focus of attention (the type and state of the current hypothesis and how it is expected to develop) is managed by the control system. Data structures passed round the loop of figure 11.5 comprise the above mentioned tokens (generated by the current schema), attached to 2D geometrical descriptions of drawing primitives. The control system reads only the tokens, while schemata are only affected by the geometric components. This means that the control system need only handle a stream of symbols; the way that this is done is described in the next section.

11.5 Control Issues

ANON's control system consists of strategy rules written in the form of an LR(1) grammar and applied by a parser generated using the Unix utility yacc (yet another compiler compiler) [262]. LR(1) parsing provides a rapid and compact method of syntactic analysis which may be applied straightforwardly to certain pattern recognition tasks [263]. It is important to stress, however, that the grammatical rules incorporated in ANON are not intended to define a legal engineering drawing in any declarative sense; rather they specify strategies by which the various components of a drawing might be recognised. It should also be stressed that we do not suggest that an entire drawing image could be interpreted by a single left to right scan producing a string generated by such a grammar. Our use of an LR(1) grammar as a control rule interpreter restricts the rules that can be applied. In this section we discuss features of the parser and its combination with schema-based token generation which mitigate these restrictions, and aspects of the drawing conversion problem which suit it to our approach.

Control rules, like string grammars, describe acceptable sequences of events; an LR parser is therefore a natural vehicle for their application. In ANON this sequence of events is the stream of tokens generated by a successful schema-driven analysis of the input image. Note that the grammatical style reflects the naturally repeating structure of drawing interpretation strategies, as was pointed out in Chapter 9. When extracting a chained line, for example, the system must locate successive line segments and gaps until some termination condition (e.g. finding a corner or white space) is satisfied. This easy mapping between control and grammar leads to a compact but powerful rule set. The example given in Section 11.6 shows how grammar rules for a chain line form a natural description of a strategy for its extraction.

In reviewing the use of logic as a representation in computer vision, Rao and Jain [264] list among its advantages the provision of a simple, expressive rule format, formal precision and interpretability and guaranteed consistency. Like logic, LR(1) parsing is a well understood, formal method which displays these properties. The existence of well established software tools based on the technique is an added bonus.

Yacc is a parser generator usually employed to create command interpreters and language compilers. It accepts an LR(1) grammar together with action code in the C language (to build parse trees for example) and generates a table-driven, stack-based, finite state machine. This calls a user-defined function to obtain the tokens to be parsed and invokes appropriate action code whenever a grammar rule is reduced (i.e. satisfied). In ANON, tokens are supplied by the current schema and action code is replaced by calls to schema-based administration functions.

Although the yacc machine operates in a single pass left to right, the in-built single token look-ahead (denoted by the 1 in LR(1)) implies a single step backtracking capability. Examining the states of the finite state machine produced by yacc shows how the stream of tokens produced by the schemata's image analysis progressively defines the structure being extracted from the drawing. A given state corresponds to several rules being active, and thus several hypotheses being supported. Note that active rules are specifying patterns of primitives which have been partially, but not entirely recovered. Subsequent image search operations may therefore either confirm or deny their existence in the drawing. As successive tokens are processed rules are reduced, hypotheses are validated and the results are incorporated into the drawing description. The number of hypotheses maintained in each state is exactly that needed to support a single step of backtracking.

The yacc software extends its basic parsing capability to cover illegal token sequences, which in the present context corresponds to failed strategies. The parser reports failure by effectively generating a special error token. Although this is introduced by the parser rather than the current schema, it can, like any other token, be incorporated in grammatical rules. As a result, the user can define situations in which failure becomes acceptable, i.e. set breakpoints beyond which any further processing cannot detract from the entity already found. This ability is essential when designing strategies for the interpretation of engineering drawings. For example, although most dimensioning is associated with nearby text, cases do arise in which numerical values are specified elsewhere in the drawing. Failure to locate text should not result in the abandonment of otherwise acceptable dimension symbols. The error token allows one to write dimensioning strategies in which text is optional, but not necessary. The insertion of such break points also means that the system can be tailored to produce partial interpretations of structures that have not conformed exactly to the model, giving ANON the robustness that is necessary to its operation.

Another aspect of uncertainty in the interpretation of drawings is the presence of points from which there are several alternative search paths, for example at the intersection of several lines. In this case the current schema not only uses its in-built priorities, but also seeks an interpretation (a token) that satisfies the strategy grammar. This is readily done by applying the yacc machine's error-checking code to each available token before returning one to the control system: erroneous tokens are rejected if one acceptable to the grammar can be generated from an alternative tracking path. Should more than one valid token be available, a schema-specific selection function is invoked.

In addition to providing ANON with a portable, efficient rule interpreter, yacc supplies a flexible user interface that allows fast prototyping of strategies. Given a suitable rule set a new control system (i.e. parser) can be compiled within minutes. Furthermore, yacc detects and reports inconsistent rules during compilation, an ability which has proved very useful during the development of the system. Although the greater modularity offered by more traditional rule languages is attractive, eliminating inconsistencies from such environments can be a long and complex task. McKeown et al [265] have reported difficulty in extending their aerial image interpretation system, SPAM, for just this reason.

With the above considerations in mind, it would seem that the reduction in generality brought about by our use of an LR(1) parser is compensated by the efficiency of the yacc machine. The customary compromise between speed and generality has been well made. Note also that the use of a high-speed rule interpreter to direct low-level processes ensures the efficiency of the overall system by keeping the amount of unnecessary image analysis to a minimum.

Mechanical drawings are man-made artifacts intended to communicate information according to a loosely predefined convention; they are read rather than perceived. It is therefore not enough simply to define the constructs that might occur in the drawing, we must determine effective strategies for their extraction and interpretation. In this situation the facilities provided by yacc make LR(1) parsing a particularly suitable method of applying control knowledge.

11.6 Entity Extraction: Chained Lines

To complement the rather general overview of ANON presented so far, we now consider its detailed operation in a particular case: the extraction of a chained line. Consider the vertical chain marking the centre of the rightmost view in the drawing of figure 11.6. Let us suppose that the first segment has been found and that a line schema is therefore current. This segment could be any one in the line; there is no context or prior knowledge available to direct the search to a chain end. On the first perceptual cycle the line schema directs image processing (figure 11.7.a) and interprets the result as a BREAK. That is, a gap in the linework is found immediately in front of the current line and a description of it, together with a BREAK token, is returned to the control module, the strategy grammar. Since no rule is as yet fully satisfied (this will not occur until the second line in the chain has been identified) no action code can be employed to modify the current schema. One of the six rules that are active is:

[Rule 1] broken_start: line BREAK LINE
{broken_start.instantiate()};

which states that a broken_start schema is recognised when a line schema detects first a gap (BREAK) and then a line segment (LINE). Once the second line has been found this rule will instantiate and make current a new broken_start schema.

In this rule format, terminals (tokens) are shown in upper case type, non-terminals in lower case. Thus line signifies the result of the earlier satisfaction of a rule causing a simple line schema to be instantiated, whereas LINE signifies a label attached to the results of examination of the image by a schema. The rule becomes active when the current schema corresponds to the non-terminal immediately to the right of the colon separator. The non-terminal to the left of the colon denotes the schema to be current after the rule has fired. Action code is delimited by curly brackets. As all ANON's action code is accessed via an appropriate schema instance we adopt the convention that the instance type is given first, separated from the function name by a full stop. In the above example, broken_start.instantiate() refers to the instantiate() function of the broken_start schema class.

At this stage, on the second cycle, the original line schema is still current. A break description is, however, now available. The current line schema uses this as a relational constraint to modify its interaction with the image analysis library, performing a single circular search at the forward end of the break (figure 11.7.b). Line tracking from the ink marks returned by this search leads to the extraction of the second line segment (figure 11.7.c) which the schema acknowledges with a LINE token. When this is received by the control module rule 1 above is reduced, causing the instantiation of a new broken_start schema which immediately becomes current.

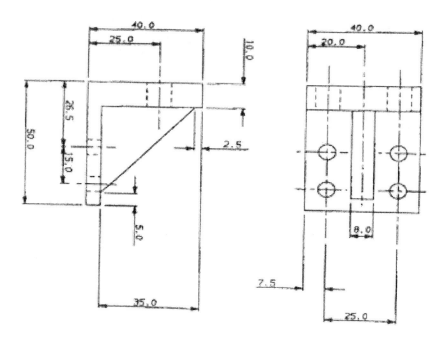

Figure 11.6. A sample drawing. Images (128 grey-levels) are taken at a resolution of 10 pixels/mm using an EG\&G RETICON line-scan camera. The field of view is approximately 20.5 by 14.5cm, resulting in a 4Mb image file. Thresholding [64] is applied for ease of display only, it should be stressed that ANON processes the full grey-level image.

The use of relational constraints between visual objects in this way is a topic that has received considerable attention in the literature. There now appears to be some consensus among the developers of schema-based systems that relations are equal in importance to the objects they constrain and should therefore be represented as schemata in their own right [248,265]. This approach has been adopted in ANON. Symbolic descriptions of the relations between drawing constructs are stored as schema instances and passed around the perceptual cycle, along with descriptive tokens, as are other graphical entities. The major difference between relational and other schemata is that the former do not contain a procedural component; instead, knowledge about how to use relations is maintained within the schema description of the affected drawing constructs.

Knowledge-directed Interpretation

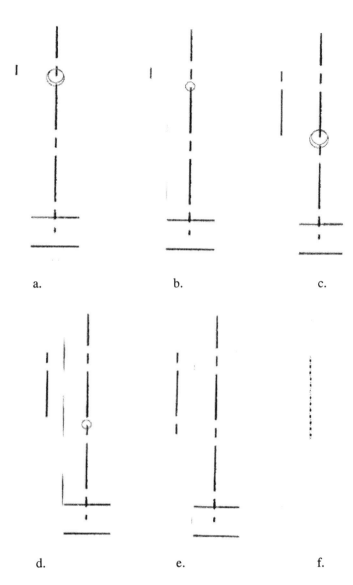

Figure 11.7. The instantiation of a chain schema. The current schema is displayed to the left of each sub-figure. Any image searches invoked are shown on the right. a) Image search under the control of the line schema. Circular patterns spread out from the end of the segment in a torch-like fashion until either black marks are discovered or the distance between the line end and the front of the beam exceeds some predefined maximum value. In this case the rear end of the next line is located by the second search. The white space between the two lines constitutes a BREAK relation. b) Image analysis by the line schema given additional knowledge of a BREAK. A circular search is performed at the front end of the gap. c) Detection of the long segment by tracking from the seed provided by b) causes instantiation of a broken_start schema which then instigates another torch-like search. d) A circular search directed by the broken_start schema after detection of the second break. The short segment

discovered by line tracking, together with the previous BREAK, allows the broken_start to be extended. e) The extended broken_start. No image analysis is required as the schema decides upon the next token by simple examination of its own contents. f) Summarised contents of the newly instantiated chain line.

The developing construct is now recognised as being a broken line of some description; whenever a broken_start schema is current it examines its contents to determine whether it is straight, curved or chained. If the broken_start can be classified as chained the token ISCHAIN is passed to the control system by the broken_start schema and the rule:

[Rule 2] chain: broken_start ISCHAIN
{chain.instantiate()};

is satisfied. This states that when a broken_start returns ISCHAIN it is specialised as a chain schema, which then becomes current. At this stage the broken_start schema examines its contents but cannot make a reliable decision and so continues the exploration. First another BREAK is discovered (figure 11.7.c), then a third line (figure 11.7.d,e). This reduces rule 3:

[Rule 3] broken_start: broken_start BREAK LINE
{broken_start.addon()};

which extends the broken_start instance. Note that the search patterns shown in figures 11.7.c,d are similar to those seen in figures 11.7.a,b. As the previously detected line schemata are sub-parts of the broken_start schema they are called upon to perform the necessary search and tracking operations. The broken_start schema now comprises three line segments joined by two break relations (figure 11.7.e). No image search is required for the next stage, the relative lengths of the lines and the regularity of their spacing is sufficient evidence to allow the broken_start schema to return ISCHAIN, reducing rule 1 above and instantiating a chain schema (figure 11.7.f).

Once a chained line has been recognised (i.e. a chain schema instantiated), further processing becomes much more directed. The chain schema can predict subsequent breaks and lines, then seek them top-down rather than bottom-up. Figure 11.8.a shows the search pattern used to locate the next (known to be long) line in the example of figure 11.7. A set of three linear search patterns, normal to the expected segment, locate ink marks which seed line tracking. The subsequent LINE, together with the predicted BREAK, satisfies rule 4:

[Rule 4] chain: chain BREAK LINE
{chain.addon()};

which extends the schema (figure 11.8.b).

Knowledge-directed Interpretation

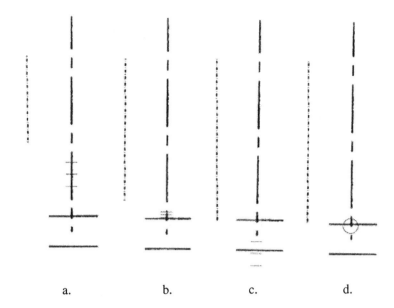

a. b. c. d.

Figure 11.8. Developing a chained line. The current schema is shown on the left of each sub-figure, the image searches it invokes to the right. A) Top down search for the next (long) line segment in the chain of Figure 11.7. b) Top down searches, performed sequentially under the control of the strategy grammar, seek the remainder of the chain's component lines. c) The last search fails to find any linework, causing the chain schema's only state variable to be set. d) With its state variable set the chain schema attempts a more tentative search pattern, which also fails to locate an acceptable continuation of the construct. Tracking in this direction is now terminated, the chain schema is inverted and extensions sought in the reverse direction.

This process continues, rule 4 being applied several times, until a predicted line cannot be found (figure 11.8.c). The current schema now returns a different token, END, indicating its failure to locate any further linework. This results in the schema's internal state variable (recall Section 11.2) being set by the rule:

[Rule 5] endchain: chain BREAK END
{chain.endset()};

No other change is made to the content of the chain schema, which remains current. With its only state variable set the chain tries a more exploratory search on the next cycle (figure 11.8.d) but fails to find an acceptable continuation. Instead it detects the JUNCTION between the chain and physical outline at the bottom of the figure, which is not legal at this point in the strategy grammar. Rule 6

[Rule 6] bichain: endchain ERROR
{chain.endclear();
chain.invert()};

is therefore reduced and the direction of the partially developed chain line inverted to allow the construct to be extended in the reverse direction. Note that the state variable is also reset as the inverted chain is no longer ending. Lines and breaks are predicted in a similar fashion during this second period of extension (not shown), which once again ends when a predicted line segment cannot be identified. Although a tentative search is tried after rule 7:

[Rule 7] endbichain: bichain BREAK END
{chain.endset()};

is reduced no further segments are discovered and tracking is terminated by rule 8:

[Rule 8] drawing: endbichain ERROR
{chain.preserve()};

which makes the completed chain (see figure 11.9) a sub-part of the initial drawing schema.

It will be noted that although only three schema instances (one broken_start, one chain and the drawing) were involved in the above example, six non-terminals have been introduced. This apparent proliferation of non-terminals is not characteristic of the system, but arises in this case because we have described only one path of the rule tree. In the full tree, a typical non-terminal represents a collection of schemata in specific states. This many-to-one mapping from hypotheses to non-terminals keeps the size of the grammar to reasonable proportions.

11.7 Top-Down and Bottom-Up Control

So far, we have shown how ANON supports the geometrical direction of image search, progressive discrimination between multiple hypotheses, and robust handling of misfit data. These have only been expounded in quite a simple case, that of a chained line. We now turn to the vital issue of system development, and the handling of more complex structures.

Chains are of a low level in that they are formed directly from image primitives and relations: line segments and breaks. Many drawing entities are more complex, being composed of higher constructs. For example, dimensioning incorporates several components including, sometimes, chains. In principle, the combination of high-level schemata to create even higher entities requires only a straightforward reapplication of the techniques discussed above, and so fits naturally into ANON. In practice, the undisciplined creation of such hierarchies could lead to an expansion of system size that would severely restrict its ultimate capabilities.

Consider two rules that contribute to ANON's interpretation of dimensioning:

[Rule 9] dimension: leader PARALLEL word
{dimension.instantiate()};

[Rule 10] dimension: word PARALLEL leader
{dimension.instantiate()};

These specify alternative strategies for the identification of annotated leader lines. The first states that a dimension schema can be instantiated and made current by a leader instance returning a PARALLEL relation, i.e. discovering something to the side and similarly oriented, and then noting that that something is a word. The second requires the word to be found first, followed by a suitably related leader.

At first sight, the need to write out separate strategies in this way might seem an unwarranted labour: it would be preferable to have a single rule describing the parallel relation between word and leader, and allow it to be satisfied in whatever order. The means to do this is not available, and for good reason. To find a word in the context of a known leader is quite different from a leader in the context of a word. In fact, developing these separate strategies is an essential step in the determination of valid and powerful rules.

It is also evident that we must be able to identify, for example, leader lines in two different circumstances. Rule 9 above provides no contextual information; the leader must be built, bottom-up, in the same way as was the broken_start schema in our earlier example. This type of entity extraction is natural to ANON. Rule 10, however, requires a leader line to be found under the control of a word schema. The word could hypothesise and test directly for a complete leader, arrowheads and all, in the same way that the chain tests for the next line segment. The difference, of course, is that a line segment is a primitive while the leader is a construct of some complexity; to take this route would be to instigate a top-down search for a high-level schema instance, an operation which requires a second method for extracting the predicted entity. Nagao et al [249] adopt this approach in their work on aerial imagery. Their program comprises a number of object-detection subsystems, similar to our schemata, which communicate via a blackboard. A given object may be detected by either of two subsystems, one operating bottom up, the other top down. For domains in which a large number of different objects may appear, such duplication results in a large and potentially unmanageable system. The problem, then, is how to examine higher-level hypotheses without constructing separate top-down test procedures.

The solution adopted in ANON is to exert top-down control only in modifying the initial search strategy, then to employ bottom-up processing. In the above example the word schema directs attention towards the region in which the leader is expected to lie, searching for a primitive which may be part of the required structure. This seeds ANON's normal, bottom-up analysis and, if a leader schema is constructed, rule 10 is reduced.

This disciplining of top down control represents an important new step. It has come to light as a consequence of applying image understanding methods to more complex structures than is usual. Without such discipline, excessive duplication of techniques is required. Of course, the solution proposed here is not the only one possible. We also expect there to be a continuing trade-off between the extent of top down control and the size of the resulting system. The solution outlined here, however, appears to provide simplicity without undue restrictions to top-down control.

11.8 Performance

The sample interpretation discussed in Section 11.6 illustrates how ANON works rather than how well it works. In the following we shall attempt to assess the system's performance. This is unfortunately a non-trivial task (discussed further and more generally in Chapter 12). First, there is no clearly defined target representation for mechanical drawings. The internal structure of a given CAD file usually depends on the manner in which the design was entered by the draughtsman. Second, huge variations can be expected in both the quality and complexity of active engineering drawings. It therefore seems unlikely that any attempt to delimit formally the range of drawings that ANON can be expected to handle would be successful.

ANON is a large and complex system. The strategy grammar contains 191 rules (out of a total allowed by yacc of 600) resulting in a 313-state parser (out of an allowed 1000). The parser recognises 122 terminals and 68 non-terminals, yacc's upper limit for each being 300. In total, ANON comprises some 20,000 lines of C code [267] and it supports fifteen distinct schema classes. Given the above considerations our evaluation of ANON must be qualitative, comparing the system's output to human segmentation of test drawings.

ANON may be run in either interactive or automatic mode, the only difference between the two being the manner in which tracking is initiated. While the system is running interactively the drawing schema prompts the user each time a new start is required. It should be stressed that the only information provided by the operator is the position and radius of a cursor used to specify an initial circular search. During fully automatic operation the bookkeeping module is used to generate suitable start positions. A search of fixed radius is then performed at each chosen point. Figure 11.9.a shows ANON's (automatic) interpretation of the drawing of figure 11.6. The initial search patterns used are shown in figure 11.9.b.

The bookkeeping module focuses attention quite well, clustering searches around text, physical outlines and dimensioning. Two points should be noted. First, no searches are performed around the middle of the chain of figures 11.7 and 11.8. This is identified early in processing and the accounts updated, directing attention elsewhere. Second, although the dark lines running down the sides of the drawing attract a number of searches, these do not generate acceptable schemata. Hence, ANON is capable of discarding any spurious starts that might be induced by image noise.

Comparison of figures 11.6 and 11.9.a is encouraging. ANON's description of the original drawing corresponds well to human interpretation and is almost complete. High-level dimensioning, chain, text and physical outline schemata account for most of the image. Some constructs are, however, missing from the system's output. Perhaps most noticeable is the absence of hidden detail (short dashed lines), missed because the pattern of initial searches was too coarse to detect its very short component segments. In some cases no attempt was made; the lines were so small that the bookkeeping module did not consider them worthy of attention. One piece of text (in the top right of the figure) was also overlooked, again due to inappropriate initial search radii.

Knowledge-directed Interpretation

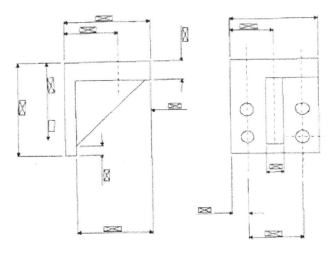

a.

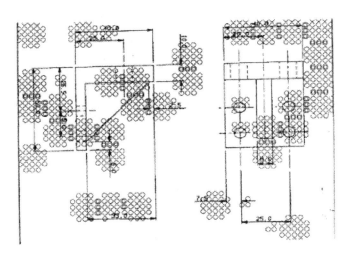

b.

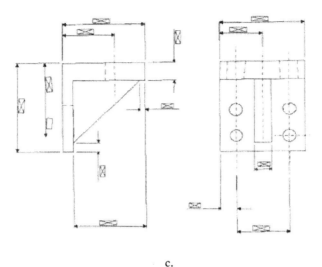

c.

Figure 11.9. a) ANON's fully automatic interpretation of the image of figure 11.6. Chains are shown as dot-dashed lines, text as rectangular boxes. The leader lines of dimension sets are marked by arrowheads. Where text is associated with dimensioning its box is labelled with a diagonal cross. Physical outlines are simply drawn as solid lines. No other schema classes are involved in this example. b) Initial, circular searches generated automatically during the interpretation of figure 11.6. c) The final interpretation, obtained after additional user-supplied starts. The description now includes previously absent text and broken lines denoting hidden detail, which are shown dashed.

While some drawing entities are missed, others are included in more than one schema. The rightmost vertical chain, for example, was found twice; first on its own as a centre line, then as part of dimensioning. The interaction between these interpretations can be seen clearly in figure 11.9.a. The leftmost chain on the right view is also found twice, though this is not obvious from figure 11.9. Two dimension sets at the bottom of the view share a common witness line, which incorporates the chain. The witness line, and hence also the chain, is therefore described by two separate dimension schemata. Only the (almost) identical representation of the witness line by these two schemata prevents the overlap from showing up in figure 11.9.a. An unfortunate consequence of overlapping schemata is that they disrupt bookkeeping; a section of physical outline was missed because, after accounting for the left chain twice, that part of the drawing did not appear to warrant further attention. Despite this problem, ANON's natural tendency to report, usually correctly, coinciding schemata is of considerable value. The initial system was later extended to explicitly seek such coincidences, allowing ANON to produce a more comprehensive description of the input drawing while reducing processing effort. These developments are considered in Section 11.9.

Many of the deficiencies of figure 11.9.a can be made up by running the system interactively after automatic interpretation. Figure 11.9.c shows the description of figure 11.6 obtained after processing nine additional starts supplied by the user. The missing text has been located, as has most of the hidden detail. Note the inclusion of

Knowledge-directed Interpretation 229

the previously absent text in a dimension schema which overlaps the one found in automatic mode. Those dashed lines that could not be added are too small to provide the context needed for schema instantiation. Similarly, it will be noted that, in figures 11.9.a and c, only one of the four horizontal centre lines in the rightmost view is represented as a chain. The others are absent, interactions with the circles having prevented ANON from obtaining sufficient clean structure to instantiate a chain schema. Omissions of this type, which are typical of schema-based systems, can only be rectified via knowledge of the type of construct expected. Hence, they cannot, as yet, be overcome in interactive mode as ANON's user can only specify where tracking should commence. Accurate positioning with an appropriate search radius does, however, give the system the opportunity to apply its existing knowledge and so provides improved output with minimal operator intervention. Significant increases in interactive performance would require the inclusion of methods similar to those described in Chapter 8.

Figure 11.10 illustrates ANON's performance given the slightly more complex drawing of figure 11.10.a. The result of automatic interpretation is given in figure 11.10.b and the final description, after ten user-supplied starts, in figure 11.10.c. Despite the limitations discussed above the description produced in automatic mode is once again quite acceptable. Most of the dimensioning and physical outline is represented, along with the larger of the chained lines. The remainder of the text, dimensioning and physical outline, missed due to the coarseness of the automatically generated starts, is successfully added in interactive mode.

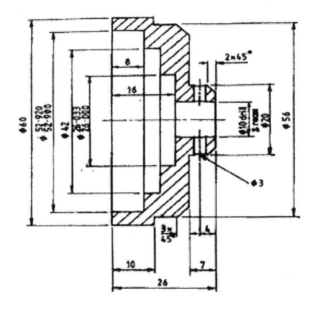

a.

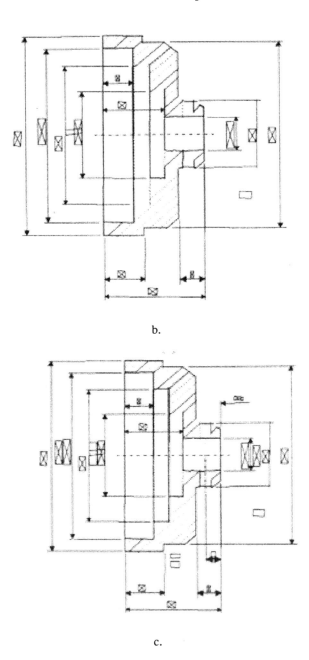

b.

c.

Figure 11.10. a) A more complex image thresholded, as in figure 11.6, for display purposes only. b) Automatically obtained interpretation of figure 11.10.a. In addition to the schema classes present in 11.9.a,c, are a number of cross-hatching schemata, drawn as sets of parallel finely dotted lines. c) Final description of figure 11.10.a, obtained after processing ten user-supplied starts. More careful positioning with an appropriate search radius allows ANON to extract the remainder of the text, dimensioning and physical outline.

Knowledge-directed Interpretation

The cross-hatched regions of figure 11.10.a are quite large and, though correctly identified, are represented (cf. figure 11.9) by several overlapping schemata. Conflicts also arise between different drawing entities. Close examination of figure 11.10.b reveals that four of the parallel lines making up the physical outline also form acceptable hatching (figure 11.11). Several of the shorter cross-hatching lines are similarly considered, wrongly, to be part of the physical outline. McDermott [268] argues that a knowledge-based system will display erroneous behaviour for either of two reasons; (1) inadequacies in its knowledge or (2) inadequacies in its problem solving behaviour. The conflicts depicted here and in figure 11.9 are clearly the result of the latter problem; their resolution requires a global view that ANON's comparatively local, schema-based operations cannot be expected to provide. A further example is seen to the far left of figure 11.9.a, where two dimension sets share a common leader line. ANON has created a single schema merging the two together. Only by considering a large part of the image can the system be expected to rectify the error. Initial steps towards the development of this capability form the subject of Section 11.9.

Figure 11.11. Areas of cross-hatching extracted from the image of figure 11.10.a. All the cross-hatching is represented, albeit in several overlapping pieces, but part of the physical outline also forms acceptable hatching.

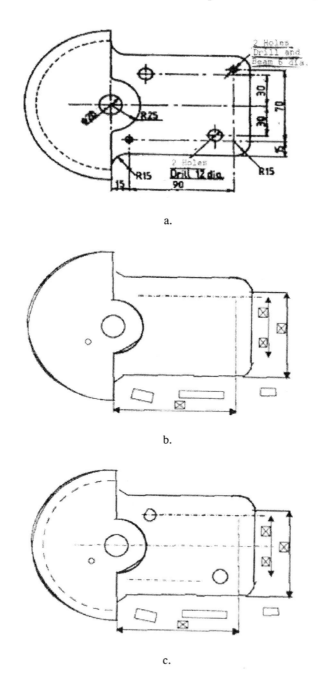

Figure 11.12. a) Input drawing image, thresholded for display, b) the result of automatic interpretation, c) the final interpretation. Note the dashed curve to the left of the figure denoting the corresponding broken arc in the drawing.

Knowledge-directed Interpretation

In contrast, the problems apparent in figure 11.12 are largely due to lack of knowledge. The drawing (figure 11.12.a) contains a number of new constructs for which ANON has neither schemata nor strategy rules; most of the text touches its underlining and there are several radial dimensions. Performance is further degraded by the use of shared leader lines in all other dimensioning. Even so, automatic interpretation (figure 11.12.b) correctly identifies most of the physical outline and as much text and dimensioning as can be expected. Interactive help (figure 11.12.c) adds more text, two of the holes, the dashed arc, centre line and part of the lower chain.

11.9 Scene Formation

11.9.1 Searching Image and Memory

ANON's basic activity is a form of classification problem solving, primitives extracted from the image are compared against schema classes in an attempt to identify an appropriate set of schema instances. While there are clearly identifiable entities that are the building blocks from which drawings are constructed, these components can be combined in very many different ways. For example, although the standard forms of dimensioning have been catalogued [232] and are readily described by schemata, they may be combined quite arbitrarily to specify the dimensions of complex objects. Under such circumstances and given that ANON is only concerned with instantiating acceptable schemata, overlapping, though similar, interpretations are unavoidable.

As noted above, overlaps may also be caused by local ambiguities. Although each schema represents a significant number of image pixels the constructs involved are still quite small relative to the drawing. Situations therefore arise in which this comparatively local processing is unable to determine the correct interpretation.

An obvious approach to these problems would be to attempt to define higher-level classes that can represent overlapping, consistent schemata while rejecting inconsistent interpretations. We feel, however, that it is naive to expect ever higher schemata to be able to capture the variety and complexity of interaction that is found in even the simpler mechanical drawings. The space of possibilities is simply too large. Instead, the extended form of ANON makes interactions between its standard schemata explicit, producing a representation more akin to a semantic net than to the collection of unconnected entities generated previously. This approach provides both a powerful representation of drawing content and a basis for the resolution of conflicts.

In its original form ANON merely displayed completed schemata to the user. Hence, as interpretation progressed, a considerable amount of information was accrued but not exploited in any way. The extended ANON makes use of this knowledge by retaining complete schemata in a drawing memory which may be examined, along with the image, on subsequent perceptual cycles. During this examination, interactions are noted between the developing schema and those already completed. The architecture of the extended system is shown in figure 11.13.

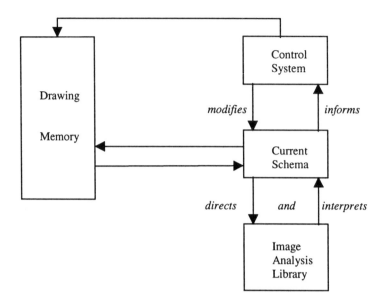

Figure 11.13. The ANON architecture extended to incorporate a drawing memory.

The question here is how can we integrate a developing visual memory into the knowledge-directed visual search performed by a system based on the perceptual cycle? Specifically, how can the sequence of events comprising ANON's perceptual cycle be altered or extended to incorporate a drawing memory? What is needed is some point of access to memory and some way of interpreting the schema(ta) returned. There is clearly no need to grow primitives as complete constructs will be supplied directly.

Consider the access problem first. As drawing memory contains only schemata extracted from the image, the latter is effectively a spatial index to the former. Moreover, schemata can only arise from inked areas of the drawing. The ink spots identified early in the cycle therefore provide a natural and efficient point at which to interface to memory. Hence, the current schema's knowledge of where to look for the next construct is used to initiate analysis of both the drawing image and the drawing memory. As several spots typically arise on each cycle a schema-specific function must decide which candidate(s) to pursue.

The hierarchical drawing memory is (naturally) searched top down; successively lower-level components are examined until an instance is encountered which the current schema interprets as worthy of consideration by the strategy grammar. Suppose, for example, that a given ink spot overlays the leader line of a previously completed dimension schema. On first examination drawing memory reveals the entire dimension set. If this is not deemed suitable the next level sub-component is considered. This is the leader instance, which comprises a line with arrowheads. Should this also be found lacking, attention passes to the next level instance, the line. Having reached a leaf of the schema tree, the search then terminates. As interpretations often overlap, any given ink spot may lie on more than one schema.

Drawing memory therefore supplies an ordered list of instances at each level, precedence being given (somewhat arbitrarily in this context) to those most recently stored.

This top-down search for appropriate instances in drawing memory neatly complements ANON's bottom-up interpretation of the drawing image: the current schema develops bottom up while, at the same time, incorporating the most significant acceptable constructs from memory. Before this can happen, however, the results of memory search must be interpreted and the control module informed of the outcome. Like the primitives extracted from the image, each instance retrieved from drawing memory must be assigned a symbolic token which is subsequently examined by the strategy grammar.

During image analysis only primitives need to be interpreted, or tokenised, by the current schema. ANON's memory search, however, may return instances of any class. This raises a combinatorial problem; each schema class needs some way of tokenising every other. In the worst case this suggests an N x N array of tokenisation functions (N=15 at present). Closer examination of ANON's behaviour, however, shows this to be unnecessary.

ANON can deal with high-level memory entities in the same way that it manages top-down image interpretation; by simply directing search to a particular region and waiting to see what the bottom-up interpretation processes produce. Recall that drawing memory is accessed through ink marks that would otherwise be used as the basis of descriptive primitives. If, having hypothesised some high-level entity, an instance of the appropriate class can be retrieved from memory via these ink spots it seems fair to assume that that instance would have been extracted from the image had bottom-up interpretation continued. Hence only class need be considered when tokenising any high-level instances returned from memory. For these constructs, at least, tokenisation is trivial.

Greater effort is needed when tokenising retrieved primitives. These lowest level entities are often detected by more exploratory search patterns which generate a number of primitives at once. Under such circumstances explicit geometrical and other tests are clearly required if the correct interpretation is to be identified. The tokenisation functions used during image interpretation can, however, be applied without change to primitives extracted from drawing memory. No extension of schema classes is therefore necessary.

In ANON, constructs retrieved from drawing memory are always given precedence over those extracted from the drawing image. Image primitives are only sought if no instances worthy of further attention are supplied by memory search. This is both efficient (retrieved instances must be at the same level or higher than those generated by image analysis) and loosely consistent with human experience. One drawback of the approach is that the system can be misled by its memory. Situations may arise in which a better construct is apparent in the image but not considered because the partially complete memory contains an acceptable, though sub-optimal, construct. A particularly interesting aspect of this work lies in exploring the extent to which such errors disrupt interpretation. At the present level of development no significant difficulties have been encountered.

An interesting feature of ANON's three-layer structure is its separation of geometric and strategic knowledge, the former being confined to schema classes, the latter to the strategy grammar. If this separation were complete one would expect the strategy rules to be applicable without change to schemata retrieved from drawing memory. There is, after all, no principled difference between instances built bottom up by the current schema and those constructed previously and now extracted from drawing memory.

This is in fact the case. All of the strategic knowledge incorporated in the original ANON is now also used to manage the combination of previously stored instances with the current schema. For example, the strategy grammar includes the rule:

[Rule 11] witness: line BREAK junction
{wit.instantiate()};

which states that a witness instance may be instantiated and made current by the location of suitably related line and junction instances. All that need be done to incorporate data returned from drawing memory is to add an extra clause:

[Rule 12] witness: WITNESS;

Together, these allow a witness to be made current either bottom up or via the extraction from memory of a complete witness instance. All subsequent rules dealing with the non-terminal 'witness', e.g.

[Rule 13] dimension: leader ABUTS witness
{dim.instantiate()};

apply regardless of how any given witness was obtained. ANON's interpretation of image and drawing content is therefore very closely integrated, not only at the search and tokenisation stages, but at the strategy level as well.

11.9.2 Coincidence Links

The practical goal here is the identification of interactions between schemata. Given the closely linked image and memory searches described above this is easily achieved. Whenever a previously stored instance is reactivated, drawing memory simply returns a copy of that instance augmented with a coincidence link to the original, which remains in memory. The copy may then be incorporated into whatever construct is developing at that time while retaining an explicit link to its previous interpretation.

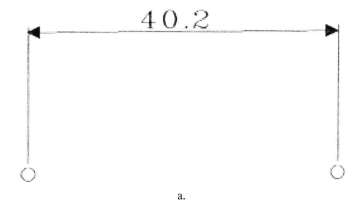

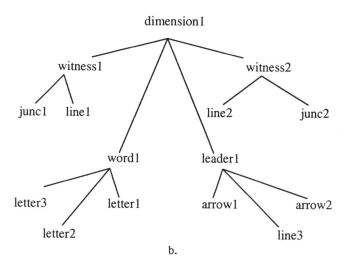

Figure 11.14. a) ANON's interpretation of the upper right dimensioning of figure 11.6. Note that the value 40.2 was not determined via character recognition, but by measuring the length of the leader line. b) the underlying schema representation.

Consider the two nested dimension sets sharing a witness line at the top left of figure 11.6. Figure 11.14.a depicts ANON's interpretation of the uppermost construct while figure 11.14.b shows the hierarchical description underlying this display. Once the dimension instance is complete it is placed in drawing memory and so becomes available to the current schema on subsequent perceptual cycles. When attention falls on the second piece of dimensioning, its leader, text and one of its witness instances are extracted from the image. The second, leftmost witness, however, is obtained, complete, from memory. Figure 11.15 shows the content of drawing memory after the second dimension has been completed. A coincidence link (figure 11.15.b) is in place between the two overlapping witness instances, which are marked bold in figure 11.15.a.

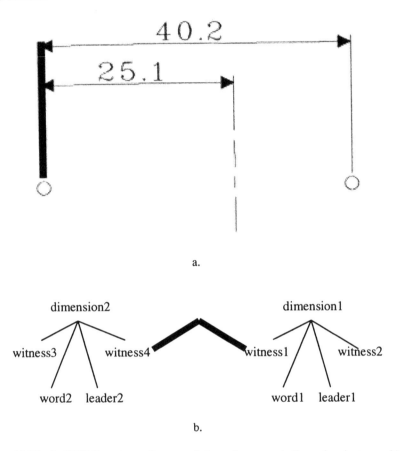

Figure 11.15. a) ANON's output after completion of a second dimension instance, b) the underlying (summarised) schema representation. Bold lines denote a coincidence link.

Figure 11.16 shows ANON's final interpretation of the drawing of figure 11.6, coinciding instances are drawn in bold. The system's description of the original drawing corresponds well to human interpretation; high-level dimensioning, chain, text and physical outline schemata account for most of the image. Note the presence of coincidence links between all shared witnesses and between the junction instances appearing in both witness and physical outline schemata (marked by black dots). This association of dimension schema with dimensioned object is particularly informative. Overlaps between chained line schemata contributing to centre line and witness instances are also visible to the right of the figure.

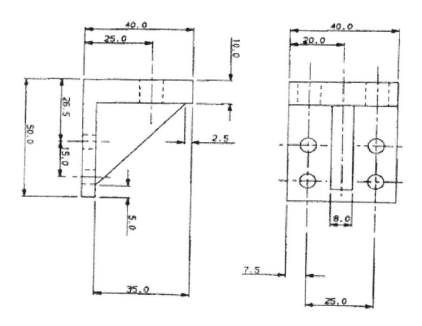

Figure 11.16. ANON's interpretation of the image of figure 11.6. Coinciding schemata are drawn bold.

11.9.3 Partial Interpretations and Schema Fusion

The above discussion deals with the effects of drawing memory on ANON's central activity: the combination of low-level schemata to create and maintain instances of higher-level constructs. The addition of this memory also provides ANON with the opportunity to extend its operation somewhat. In particular, the system is now able to fuse together overlapping instances of the same class, a facility that is invaluable given the variation found in mechanical drawings.

Suppose, for example, that a developing cross-hatching instance overlaps one already held in memory. ANON now combines these two schemata to create a single instance representing the full extent of the hatched region. This is in contrast to the two independent (but overlapping) schemata reported by the original version of the system. As hatching accounts for arbitrary regions of the drawing it is impractical to expect entire hatched areas to be extracted, reliably, in one pass. The level of strategic knowledge required would seem to be prohibitive. ANON's cross-hatching strategies are therefore limited to seeking convex regions which may be fused together later to form more general hatched areas.

This ability to locate entities piecemeal and assemble a more complete description in memory greatly improves ANON's final interpretation; schema fusion leads to fewer instances, each covering a larger portion of the image. The system is currently able to fuse hatching, physical outlines and centre lines.

ANON deals with partial interpretations in similar fashion. As with all schema- or frame-based systems, situations arise in which a given instance is created but cannot be completed. Such partially filled schemata can represent a significant processing effort but are often discarded immediately. ANON may now defer its rejection of (some) partial instances until processing terminates. Those partly filled schemata that are large enough to warrant further attention (e.g. dimensions lacking only text) are held in drawing memory, but labelled as partial. This gives the system every opportunity to incorporate these constructs into other, complete, interpretations, taking advantage of the earlier effort. Any schemata which remain partial on termination are discarded.

11.10 ANON in Context

It is useful at this stage to situate ANON with respect to some other knowledge-directed image analysis systems. Two initial points should be made regarding general differences between ANON and other systems. First, ours is an expert task; reading complex drawings requires training. ANON is not, therefore, intended to be a model for general vision. Second, as drawings are naturally read sequentially, strategic knowledge is of particular importance; hence ANON's reliance on the control rules made explicit by its strategy grammar.

Although parsing is a standard tool in syntactic pattern recognition, its use as a control mechanism is less common. Tropf and Walter [269], however, use an Augmented Transition Network (ATN) to specify strategies for the recognition of modelled industrial parts. While their system works well, its reliance on procedural knowledge seems excessive given the task at hand. The ATN specifies the order in which lines and corners extracted from an image may be matched to corresponding features of a geometric model, and therefore plays a role similar to that of ANON's grammar. However, though they refer to their approach as analysis-by-synthesis [270,271] the strategies embedded in the ATN bear no relation to the manner in which their images were either created or intended to be perceived. Hence this type of procedural knowledge seems more suited to the interpretation of communicative documents than to images of physical objects.

At an abstract level, the non-terminals of our strategy grammar form the nodes of a (directed) graph in which each arc represents a relation or predicate. This corresponds closely to Mulder's [250] discrimination graphs. The major difference here is that, while Mulder's graphs must be acyclic, our LR(1) parser can handle cyclic structures. As interpretation proceeds the graph is searched, under the control of the parser, for paths from the root to acceptable goal nodes. At each step some given predicate or relation must be satisfied if attention is to pass from one node to the next.

Despite the advantages afforded by the cycle of perception architecture, few computer vision systems have taken this approach explicitly. Glicksman's [272] MISSEE is a notable exception. This is also schema driven, its goal being to produce a semantic network of schemata describing an aerial image of a small urban scene. MISSEE's schemata communicate directly, so control knowledge and schemata are not distinct but integrated within the semantic net. This integration raises questions regarding the expandability of the system [272]. A further difference between ANON and MISSEE is the latter's use of a pre-segmented image; all low-level processing is completed before interpretation commences. This separation of image description from higher-level interpretation is commonplace, even in systems employing schemata. VISIONS [257], SPAM [251], MAPSEE [248,250] and the systems due to Sakai et al [252] and Nagao et al [249] all operate on pre-processed images.

GOLDIE [254] does use schemata to direct image analysis. Moreover, like ANON, its low-level routines are maintained as a separate library of image processing tools. GOLDIE's schemata, however, do not represent expected constructs. Instead, they describe strategies by which different types of image analysis task (region segmentation, line extraction, collinear line grouping, etc.) may be achieved and are invoked in response to goals posted by higher processes.

Matsuyama and Hwang's [253] SIGMA both represents its knowledge of expected objects as schemata and combines low-level processing with interpretation. SIGMA's schemata do not, however, control image analysis directly. This task is delegated to a Low-Level Vision Expert (LLVE) which uses knowledge of the available image processing operators (cf. [273]) to satisfy goals posted (cf. GOLDIE) by the higher-level components of the system. New goals are posted as schemata are instantiated. At the top level, SIGMA's operation is governed, not by explicit strategic knowledge but by a spatial reasoning module which attempts to find consistent sets of schemata.

In its present form ANON's schema-based operations provide a layer between the grey-level image and the higher-level, more global processes needed to integrate pieces of drawing into a coherent whole. A similar approach is adopted in SPAM [251] and the later versions of MAPSEE [248,250]. In both systems schemata, called functional areas in [251], are created and filled before some more global process (constraint propagation in the case of MAPSEE, higher -evel rules in SPAM) resolves any conflict between competing interpretations. Although its schemata appear to be more powerful, ANON currently lacks the necessary higher-level component.

11.11 Discussion

So far, ANON has been applied to images of piece-part drawings. It is hoped, however, that much of the code will prove to be reusable in other engineering domains. The image analysis library, based as it is on sequential tracking of linework, should be applicable to most types of drawing and many of the constructs represented by ANON's schemata are commonly used elsewhere. The system's modular structure makes it easily extendible, new image processing techniques, schemata and/or control rules may be incorporated as required. Software tools provided by yacc ease the addition of the latter feature considerably, allowing compile-time detection of inconsistent strategies.

While our evaluation of ANON has been rather informal, the results obtained to date are encouraging. The examples considered above demonstrate the system's ability to provide high-level descriptions which correspond well to human segmentation. Many of the deficiencies of the automatically produced representations are the result of inappropriate initial searches.

The successful incorporation of drawing memory provides further evidence of ANON's interpretive power. One can characterise drawing interpretation as the search for acceptable structures in the presence of distractions; when pursuing a given hypothesis it is important not to be led astray by the extraneous linework and symbols which typically disrupt the target construct. While supplying much useful information, drawing memory is also a clear source of additional distractions. That this increase in potentially misleading data does not cause ANON to fail unduly confirms the success of the basic system design.

Despite this, further extensions are necessary. Considerable work remains in identifying suitable methods for the resolution of conflicting interpretations. ANON's explicit representation of overlapping schemata, however, represents a significant step towards the construction of the scene constraint graph [cf. 248] which might form the basis of such work.

It should also be emphasised that we do not consider that the current ANON system would provide useful fully automatic conversion of drawings to CAD format. It would, however, provide a valuable automatic element to combine with a manual interface, and, more importantly, does provide a methodology for the future development of an increasingly automatic system.

Chapter 12
Current Issues and Future Developments

12.1 Higher-Level and 3D Representations

12.1.1 Resolving Inconsistency

For many, entity extraction is the primary goal of line drawing interpretation; the majority of the published literature in the field is concerned with processes contributing to or preceding the recognition of drawing entities. However, like vectorisation and primitive extraction, entity recognition may be seen as a step towards an even higher-level interpretation. The representation produced at this higher level provides a global view in which locally recognised entities are combined to provide a description of the entire drawing. Our own initial steps in this direction were described in Chapter 11.

The output of an entity extraction process can be viewed a number of ways. It may be a set of hierarchically grouped primitives and elements whose structure reflects that of the entity they represent; a flat set of primitives, each of which is assigned a label describing its role in some entity, or a combination of these two extremes. Whatever the form of its output, entity extraction is sufficiently difficult that a perfect solution is unlikely to be obtained. In real drawings entities touch, overlap and otherwise interact with each other in a variety of ways. These interactions are almost certain to disrupt the extraction process. Even if they do not, the entities produced are unlikely to be completely spatially independent of one another. Primitives may be given the same label by different entities, for example, in dimensioning more than one leader line often connects with a single witness line (Section 11.9). Overlaps such as this are likely to be acceptable and represent economic use of primitives/pen strokes by the draughtsman. Primitives may be assigned different labels by different entities. Such cases are more likely to arise from conflicts between entities arising from ambiguities in the drawing and/or the knowledge used to interpret it. Some such labellings may, however, be acceptable; for example, a line segment could be both part of a centre line and appear as a witness line in a dimension set. Primitives may even be assigned different labels by the same entity, though it seems likely that ambiguity at this level would be resolved during entity extraction.

It would appear, then, that if a globally consistent drawing interpretation is to be obtained some form of ambiguity/conflict resolution process must either be incorporated into, or follow, entity extraction. Little attention has been paid to this process by the drawing understanding community. A notable exception is the OO-

Mudams system of Wu and Sakauchi [274]. This uses the Truth Maintenance System proposed by Doyle [275] to detect and resolve inconsistencies in an object-oriented interpretation of an input drawing. Truth maintenance is a potentially powerful technique but can, however, be very expensive computationally. As described in Section 11.9, later versions of ANON went some way towards identifying inconsistencies, but made no attempt to remedy, or even investigate, those found. The consistency problem has received more attention in the wider image understanding community. While interpreting aerial images of airports, SPAM [251,265] uses rules encapsulating the likely relationships between entities (hangars, runways, control buildings, etc.) to both guide entity extraction and reject inconsistent interpretations. A similar approach is employed in the higher-level components of SIGMA [255] which exploits constraints on the likely layout of entities (houses, roads, etc.) extracted from aerial images of suburban scenes. In sketch-map interpretation, Mackworth et al's various MAPSEE programs [248,250] comprise frame-based entity extraction followed by a co-operative process which identifies, and attempts to resolve, conflicts between reported entities (rivers, mountains, bridges, etc.). The schema system of Draper et al [276] employs a distributed conflict detection and resolution system. When schemata compete for some image area either of two mechanisms may be brought to bear. When predictable conflicts arise, dedicated recognition processes are invoked to extract further information from the image and resolve the situation. If no dedicated process has been specified, confidence values associated with the competing hypotheses are used to make a decision. Fuzzy logic has been used to resolve conflicts by Agazzani et al [277] and Dellepiane et al [278]. Both systems resolve conflicts by re-segmenting the disputed areas using modified segmentation parameters.

Implicit in this work is a view of image interpretation systems as sets of interacting layers. Primitives may be identified by an initial context-free process or via interaction between that process and the higher, entity extraction level. Similarly, entity extraction may proceed purely on the basis of the data supplied from below and knowledge embodied in the entity extraction layer or via interaction/communication with a higher conflict resolution layer. Whether or not this architectural picture persists, some form of conflict resolution strategy and engine is necessary if the full benefit of current entity recognition systems and techniques is to be gained. The topic is currently under-researched in the line drawing interpretation community; partly because other problems remain to be solved and partly because conflict resolution is a problem more closely related to artificial intelligence than to the pattern recognition which has underpinned most previous work on line drawing interpretation.

12.1.2 Editing and Parameterisation

Regardless of whether entities are formed by grouping together or simply labelling geometric primitives, most entity representations rely heavily on the underlying vectors to provide a geometric description of the drawing. One problem with vector models is that the length and width of individual primitives and the relationships between them (points of intersection, angular and distance measures, etc.) do not accurately reflect the geometry of the input drawing. This greatly limits the direct use of both vector and entity descriptions in geographical information and CAD

Current Issues and Future Developments 245

systems. The problem is widely recognised; many workers include an explicit editing and parameterisation step (figure 2.16/10.1) in their processing schemes.

Inaccuracies arise in vector and entity descriptions for a number of reasons. Errors may have been made on the initial drawing, for example. More generally, few engineering documents are drawn to scale with any accuracy. Indeed, given that many describe objects to 1/10 and even 1/100 mm, it is not reasonable to expect accurate scaling. Dimensioning and other annotations are intended to compensate for this. Many geometric properties are not stated explicitly, however, but implied. In figure 12.1, for example, segment 1 is clearly intended to be perpendicular to segments 2 and 3, though this is not stipulated on the drawing. Image noise and other distortions introduced when scanning poor quality drawings cause further problems. There is no guarantee that after vectorisation segments 1-3 will appear in precisely the correct configuration. It is similarly unlikely that segments 3 and 4 will be represented by exactly parallel line segments and segments 5 and 6 by collinear vectors. It may even be that the end points of, for example, the primitives representing segments 4 and 8 or 8 and 5 do not coincide.

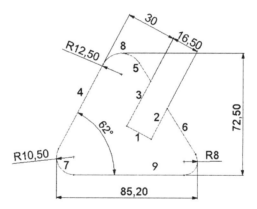

Figure 12.1. Implied geometric relations in an engineering drawing, see text for details.

A need therefore arises for a suitable means of editing entity and vector descriptions, a process generally referred to as parameterisation (the term employed here) but sometimes described as the resolution of inconsistency [279]. Current CAD systems (e.g. SolidWorks and PRO/Engineer) often incorporate parameterisation tools which can be applied to user-prepared drawing representations. The parametrrisation of automatically vectorised drawing images, however, has not been widely considered in the literature.

Broadly speaking, a drawing representation may be edited in either of two ways. In the first the model is immersed in a graphic editor, e.g. AutoCAD, that supports the removal and reconstruction of erroneous primitives. This method can be used to correct distortions introduced during scanning and vectorisation, but the process is often rather time-consuming. Suppose, for example, that the entity recognition/vectorisation system decides that in figure 12.1 the vector corresponding to segment 4 should intersect arcs 7 and 8. In enforcing these relationships it may reject the previously extracted segment 4 and replace it with a new vector that touches segments 7 and 8 in the appropriate places. This will probably affect segment 4's relationship to other primitives. It might, for example, change the angle between segments 4 and 9, which may have to be corrected. If the system later modifies the radius R12,50, without moving the centre of the arc, the angle between segments 4 and 9 will again change and it will again be necessary to rebuild segment 4 to make it intersect arcs 7 and 8.

In the worst case it may take almost as long to complete this process as it would to reproduce the paper drawing within the graphics editor, by hand, from the very beginning. This is perhaps not so surprising; systems like AutoCAD were designed to support the creation of new drawings. As a result they require the position, orientation, size, etc. of each drawing component to be specified precisely. Until an automatic drawing interpretation process can provide reasonable quantities of such information reliably it will do little to ease the task of creating CAD drawings via AutoCAD-style tools.

In the second approach, the drawing representation is thought of as a sketch to be updated. The production an accurate drawing representation therefore becomes a search for the best parameterisation of a set of models of geometrical objects, those models being provided by an automatic interpretation process. Quantitative constraints (required sizes, angles, etc.) are added to the sketch and a correct model built automatically. Constraints such as parallelism, collinearity, etc. between primitives may be recognised automatically or specified in interactive mode. Automatically identified constraints may be corrected manually. Once the required relations have been finalised, they are applied to, and their effects propagated across, the drawing model, updating and correcting the geometric properties of the initial vector/entity description. This facility is provided by various systems, for example GCAD [280], SolidWorks, PRO/Engineer and DesignPost Drafting.

If we consider the whole process of converting paper-based drawing projections to 3D objects a question arises: when should geometric primitives be matched with the corresponding dimensions? There are two possibilities [280]:

1. After vectorisation, but before 3D model reconstruction. This provides an opportunity to apply well-developed and successful methods for the parameterisation of 2D object models.

2. After reconstruction of a rough 3D model created using approximate drawing co-ordinates. Dimensions and relations established within the 3D model may then be propagated more widely across the drawing. The drawback of this approach is the poor level of development of current methods for the parameterisation of 3D models.

To the designer of a parameterisation system, the main problem with automatically derived line drawing descriptions is the high probability of incomplete dimensional information. As noted above, many dimensions are implied rather than stated explicitly. The problem is compounded when the drawing comprises several projections or sections; to make use of all the dimensional information it may be necessary to form quite complex projectional links. Some auxiliary lines (marking

Current Issues and Future Developments

for example, the centres of circles and axial lines) are by convention never dimensioned. Moreover, the available dimensional information cannot always be taken at face value; conflicts can arise between textual/numerical data and linework. Figure 12.2, for example, shows a right-angled triangle with internal angles of 45 degrees. The annotation, however, gives one angle as 50 degrees. If the written value is accepted the positions of the corners of the triangle must be modified and vice-versa.

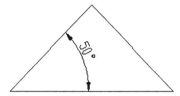

Figure 12.2. A simple conflict between linework and annotation.

The parameterisation tools built into existing CAD/graphics systems are constructed on a variety of principles, each of which imposes different restrictions on their application to automatically vectorised drawings. A linear, sequential model construction method is realised in GCAD [281]. A non-directional graph is constructed in which nodes correspond to structural elements (segments, circles, points) and arcs correspond to known relationships (intersection, parallelism, fixed distance, etc.) between them. The graph is then arranged so that enough arcs enter each node to define, unequivocally, the corresponding structural element. The method is subject to the following restrictions:

1. it can only be applied when the object considered can be constructed, consistently, element by element;

2. there are no contradictions in the available dimensioning.

Sequential construction of drawing objects is almost always possible, the second constraint is more frequently violated. The algorithm has the advantage, however, of being able to detect insufficient or superfluous dimensioning of the input drawing.

PRO/Engineer also requires clear and consistent dimensioning to be input before parameterisation can begin. The system supports the use of various dimensioning systems, including those for which it is not necessary to find sequential chains of constraints. SolidWorks also supports a variety of dimensioning schemes and does not require an initial, completely dimensioned model.

Figure 12.3 shows the result of vectorising a slightly simplified version of the engineering drawing first seen as figure 10.2. The vectors are recorded in DXF format and imported into SolidWorks. Here it is edited with the aim of eliminating any distortions introduced during scanning and automatic vectorisation. Editing comprises:

- removal and replacement of deformed elements;

- removal and replacement of all automatically generated dimensioning.

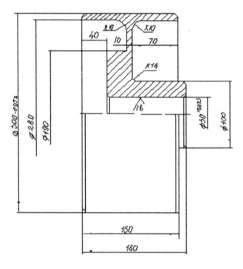

Figure 12.3. A vectorised engineering drawing (see text).

The latter operation is necessary because all sizes recorded in SolidWorks must be associated with the corresponding linework. Without this, parameterisation is impossible. SolidWorks can now identify relations implicit in the given drawing; horizontal and vertical position, intersection, etc. Connectivity between drawing elements and the location of their common points is established automatically during input of a DXF file into SolidWorks. After recalculation of the model within the system, a new model is created whose parameters correspond to sizes given explicitly, and relations given implicitly, in the initial drawing. This model is shown in figure 12.4. Figure 12.5 shows the same drawing after recalculation of its dimensions. The final drawing model can be used in subsequent computer-aided design tasks. It might be necessary, for example, to produce a similar drawing but vary some dimensions. Given the presence of a relational model of the drawing (i.e. a model comprising a set of structural elements and their relations) this task is simply achieved in SolidWorks.

The main obstacle to effective, automatic parameterisation of automatically generated vector and entity descripions is the lack of complete and consistent dimension information. Advances in entity recognition methods may help, but it seems highly unlikely that the necessary information will be provided by any automatic drawing interpretation system until significant progress has been made towards the principled resolution of inconsistencies (Section 12.1.1). In the meantime, the construction and use of interactive editing tools appears to provide the best route to geometrically correct drawing interpretations. It should be noted, however, that the construction of powerful, easy to use parameterisation toolkits and environments will require at least some understanding of, and possibly even a partial solution to, the inconsistency problem. Even if the operator is to make all the necessary corrections and modifications, the tools he/she is to use must be capable of

Current Issues and Future Developments 249

dealing effectively with the range of problems expected to arise. The scope of these tools will be hard to specify without reference to the inconsistency problem. Moreover, specifying the interaction required between human operator and parameterisation environment may also be problematic unless some knowledge is available of the type of distortions and inconsistencies expected.

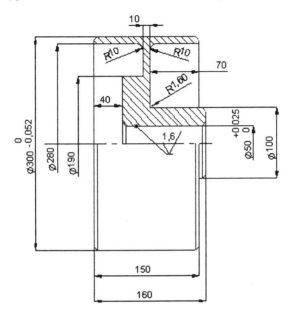

Figure 12.4. An edited and parameterised version of the drawing of figure 12.3.

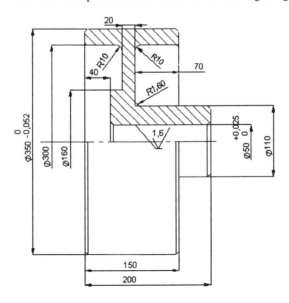

Figure 12.5. The drawing of figure 12.3 after the recalculation of dimensions.

12.1.3 3D Reconstruction

The final goal of engineering drawing interpretation is the construction of a 3D model of the drawn object(s). A robust ability to generate such models from paper drawings would provide the designer with a powerful, alternative method of access to the shape of a proposed object [282-284]. Assuming that the input drawing has been interpreted to the level of CAD primitives, 3D reconstruction is achieved by the analysis of multiple projections. The recovery of 3D models has been the subject of increasing study over the last decade, as image interpretation techniques have become more mature and successful. Although further work is required, current systems and techniques begin to provide a sound basis for the transformation of technical projections into 3D object models. Many CAD systems, e.g. AutoCAD, already incorporate some form of model building tool.

A typical engineering drawing comprises some subset of six orthogonal views: front, top, right, left, back and bottom. Further, auxiliary views are often added to improve representation of oblique faces. Three orthogonal projections, however, form the minimal input required for 3D reconstruction.

Numerous purely geometric reconstruction methods have been proposed, the basic idea being to look for sets of consistent matches between features extracted from three projections of an object [285]. One group of algorithms, based on the concept of "fleshing out" projections, was proposed in [286]. These are based on the reconstruction of a wire-frame model by matching vertices and edges between projections; the faces of the object are then obtained by propagating constraints across the wire frame. Many variations on this theme have been developed. A second group of algorithms is volume-oriented, the idea here is first to find 3D subparts, then to combine them to build the complete object.

As Tombre [285] notes, all these methods share two limitations which narrow their domain of application:

1. The data on which they work has to be perfect, i.e. they require a clean, idealised set of projections, uncluttered by annotation and free of noise and uncertainty. This is, at present, impossible to obtain reliably via automatic interpretation of images of real drawings.

2. They only consider the geometric component of the drawing. A large part of most drawings is, however, more symbolic than geometric. The large amounts of valuable information carried by the symbolic component cannot be exploited by these methods.

The first limitation might be overcome, or at least eased significantly, by the development of more reliable automatic techniques. Interactive methods could also be incorporated to improve the quality of the input data. The second limitation is perhaps more difficult to deal with, though it might be overcome by applying knowledge-based methods (see Section 12.2 below). Some potential solutions are considered in [40,287-295].

To provide an example of current methods of 3D reconstruction from multiple projections we now briefly describe the technique proposed in [296]. This makes several assumptions. First, the methods involved are tuned to selected types of machine-building drawing, used in machine-building attachments (jigs, fixtures, etc.) and depicting, at most, second-order surfaces. Drawing scale is also assumed to be one of a standard set; 1:1, 1:2, 1:4, 1:5, 1:10, 2:1, 4:1, 5:1 or 10:1. Second, it is assumed that each object can be represented by a combination of members of a library of standard volumetric primitives (generalised cylinders, prisms, nuts, holes and others defined by the user). Finally, it is believed that any successful reconstruction technology will require the synthesis of automatic and interactive operations. In particular, the user should provide semantic information about each set of orthogonal projections, but the reconstruction software should determine the object's representation in terms of volumetric primitives.

The process is as follows, beginning with an edited and parameterised entity level drawing description:

1. Interactive specification of drawing areas corresponding to standard projection views. This includes determination of the co-ordinates of rectangles bounding each view.
2. Extraction of closed contours from the entities recognised in each view and the identification of inter-projection links between them.
3. Determination of a sketch contour for each view.
4. Extrusion, normal to the plane of the projection, of all the contours in each view. The height of the extrusion is equal to the longest corresponding side of the bounding rectangles of the other projections.
5. Removal, in interactive mode, of internal contours from each sketch.
6. Alignment of extruded views with the top view, taking into account the vertices of the bounding rectangles.
7. Boolean intersection of extruded projection views. The result is a solid model of the input object.
8. Projection of the solid model back onto the standard projections and comparison with views provided by the input drawing.

The result of applying this algorithm to the three projections shown in figure 12.6 is given in figure 12.7.

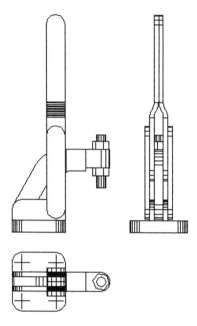

Figure.12.6. Three projections of a 3D object.

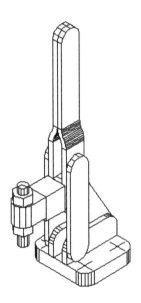

Figure 12.7. The 3D object reconstructed from the projections shown in figure 12.6

Automatic parameterisation requires complete and consistent dimensioning. Automatic reconstruction of 3D models is only possible if the available dimension information is complete, consistent and accurate. Just as some knowledge of the likely inconsistencies between vectors and entities is needed to support work on parameterisation, at least a partial understanding of and/or solution to the parameterisation problem is needed if truly robust 3D reconstruction schemes are to be developed. It may be, however, that methods will arise which combine 3D reconstruction with parameterisation, probably using the former to influence and/or control the latter. One might argue, for instance, that the primary aim of parameterisation is to support 3D reconstruction and, conversely, that a given 3D interpretation is only acceptable if it can be used to generate a consistent set of 2D entity parameters. Given this approach the two problems are intimately linked and should be solved together, perhaps by a single process. Similar arguments and combined processes have been put forward in other areas of image understanding.

The integration of parameterisation with 3D reconstruction is just one example of the interactions between operations that become possible at the higher levels of drawing interpretation. The combination of dimensioning and linework during 3D reconstruction leads to the correction of errors and, hence, to precisely dimensioned geometry, which can in turn be matched between views. The results of this matching must be combined with functional and syntactic information to construct a true 3D CAD model. Access to such a model then provides the opportunity to close the loop; revisiting entity recognition to achieve reliable identification of the items represented on the original drawing. All of these possibilities, however, require careful and appropriate representation and use of domain knowledge.

12.2 Exploiting Domain Knowledge

The value, both actual and potential, of domain knowledge to line drawing interpretation systems has been a recurring theme of this book. In papers published on topics from binarisation, through entity extraction, to 3D reconstruction one will find frequent references to the knowledge being exploited and/or the additional knowledge that will be exploited in the future. Most of the techniques that employ significant amounts of a priori knowledge do so either by hard-coding it into the algorithms they apply or by adopting a blackboard-style architecture (Chapter 2). The former approach results in potentially powerful, but generally inflexible, solutions. The latter often leads to flexible, but unstructured, difficult to maintain and inefficient systems. The power of the blackboard architecture is its generality; it is a general computational engine which can be used to implement almost anything. The down side of this is that it imposes no constraints or guidelines on the use of the knowledge it embodies.

Despite the common acceptance of the importance of domain knowledge to drawing interpretation, comparatively few projects (but see Chapter 11 and [297]) have attempted to determine what type of knowledge-based system architecture(s) are suited to the drawing interpretation task. If the oft-stated potential of knowledge-based techniques is ever to be realised, further research in this area is required. It is, of course, hard, if not impossible, to design an appropriate knowledge-based system architecture without first being clear about exactly what the role of the available knowledge is to be.

In terms of current knowledge-based image processing we can characterise the entity extraction problem as comprising two kinds of search [298]. The first is the syntactical pattern recognition (SPR) approach: isolate a small subset of the image data (corresponding to perhaps a single symbol) and then search for that pattern amongst a large set of possible drawing entities defined by structural rules. In this type of system the input data is first segmented, then each segment classified. The second type takes a model-matching approach, searching through a large set of image data for instances of a simple model description. Here segmentation only occurs as a side-effect of model recognition. The two approaches can be combined: in circuit diagram interpretation, for example, connecting lines are found by matching, then the remaining symbols are segmented out and classified using SPR. Knowledge-based image interpretation thus brings two search processes into drawing interpretation. Since both seek incomplete structures amongst noisy data, both may be exponential in time. Clearly, the only problems that are tractable will be ones where the size of search can be restrained by (knowledge of) special circumstances.

One practical approach to knowledge-based extraction, therefore, is to decide on a key entity (e.g. roads, thick lines, connection lines in schematics) which is widespread and can be reliably found from a few cues, and extract that on a best first basis. For example, when extracting map isolines [299], we start the recognition process from longer isoline parts, then try to extend them before extracting surrounding objects. With this method we are concentrating on one object at a time, and using it as the foundation for the interpretation. In this example the technique is particularly efficient because the matching search concerns a widespread and well-defined class of object, and no SPR search is used.

One can see from the above that successful techniques partition the interpretation problem up so that the computational cost is restrained. This partitioning must be done according to the demands and resources of the situation. Consider again the circuit diagram problem. First a primitive is found which can be successfully classified on a unary test (the connecting line). This then leaves the connected symbols as isolated small groups of other primitives. The same approach may not succeed in other drawing interpretation situations, because primitives cannot be so readily classified, and the resulting classification does not partition up the problem so neatly and/or completely. However, if we look at these situations we can see that the way in which the primitives used to represent the drawing relate to the sets of entities to which they belong is very important. The availability of a key entity which can be identified from a primitive, or simple combination of primitives, and which also has the property of dividing up the drawing into independently processable parts, is the basis for many working systems.

One effective way in which domain knowledge may be embedded in line drawing interpretation systems, then, is by domain-specific selection or design of descriptive primitives which partition the problem appropriately. Note that it is not enough to simply partition the problem; the partitions created must themselves be amenable to efficient search, perhaps by SPR over a small set of patterns or further, similar partitioning. Choice of representation scheme does not, therefore, just affect the way in which a drawing is described; it can have a profound effect on the ease with which a description can be obtained.

We see from the above that, despite the accepted need for knowledge-based line drawing interpretation, mere application of the programming techniques employed in knowledge-based or expert system development is no guarantee of success. A complex network of thinned lines is not susceptible to entity extraction by an elaborate rule base, due to the resulting computational complexity. Continued refinement of descriptive primitives and methods for their extraction is essential; knowledge-based approaches have no chance of success unless the data structures on which they operate make the interpretation problem tractable.

Given such tools, effective control of the entity extraction process is possible by establishing promising contexts [297,300] via the identification of entities with distinctive primitives or with distinctive relations between primitives (crossing, crossing under defined angle, joining, etc.). Further, restricted search is then performed within these contexts. If the identified contexts are powerful enough, both error rates and susceptibility to noise and clutter will be reduced. Working within a well-defined context provides opportunities to employ top-down interpretation strategies with a reasonable expectation of success.

It might be thought that the above observations constitute an artificial restriction in system design. However, if we look at recent work in model matching we see that the same notions are being applied in quite general approaches to the search problem. It is now recognised that candidate matches must be evaluated, and that the best combinations of them must be examined first if the search is to terminate in a reasonably short time. The methods that have worked in drawing interpretation are exactly those which avoid being halted by excessive complexity.

Successful use of knowledge depends on our ability to formulate rules which describe the entities concerned. Such rules are straightforward in the case of an isolated entity with no constraints of physical placement. Objects such as broken lines, roads, arcs, etc. obey rules of primitive size, orientation, etc. which enable them to be found by local sequential techniques or global accumulator techniques. When these objects meet other objects, however, simple rules no longer hold. More precisely, any rule set capable of describing the situations that may arise will have to be impractically large. This is also the case when the object placements denote a physical shape; the space of possible physical shapes is too large to be captured by a usable rule base. In general, then, we see that the rules only apply when the objects have a purely diagrammatic role; when they are (as is always the case in practice) also graphical, the rules break down.

It is tempting to extend this observation to propose that knowledge of the diagrammatic (or symbolic) structure of objects can be readily used in automatic extraction, but that the knowledge of physical reality or of the significance of combinations of entities is harder to use. Some progress on the use of spatial relations has been made, however [200,300], and the interaction between these relations and the symbolic function of graphical elements is an interesting area of research. More work also needs to be done on the accuracy and effectiveness of criteria for extraction of cartographic entities by investigating object spatial relations for particular map types.

In the ANON system (Chapter 11), an attempt is made to consider more than one entity as the search proceeds, and the set of entities considered is reduced as evidence from successive primitives is combined. The design of this system, too, is based on creating primitives, and combinations of primitives, which have powerful classification properties. This, then, systematically extends the established technique for drawing interpretation. The ANON methodology at its present level of development, however, is limited by two assumptions. One is that the resulting classification is sufficiently precise without additional SPR to define a context for avoiding noise and distractions and completing the interpretation. The other is that more global constraints are not required to resolve the classification problem (Section 12.1.1).

Constraints and conflicts between entities are unlikely to be handled using the model matching/SPR techniques employed in entity extraction; during scene formation we are not searching for a particular drawing amongst the set of all possible ones. It may, however, be possible to implement conflict resolution as the application of an inference engine to a knowledge base, for two reasons. First, the problem is comparatively well-defined over a range of drawing types. Second, the data to be input to such a component is also fairly stable as a definable set of primitives. It seems reasonable to expect, then, that conflict resolution might be achieved by application of a (fairly) domain independent conflict resolution engine to a (fairly) domain independent data structure. Most knowledge-based approaches have, however, been directed at entity extraction, so this hypothesis remains to be properly tested.

12.3 The Role of the Operator

Commercially, two approaches have been taken to line drawing interpretation [301]. The first is to suggest manual input, but with CAD-style support. This is similar to redesigning at a CAD terminal, but with the aid of an intelligent system which incorporates an image of the drawing, knowledge about the drawing and a sophisticated interface allowing the system to question the user regarding line types, their intersections and geometrical relations. A related approach [302] is to allow the user to sketch his/her design on some type of electronic drawing board, providing auxiliary information verbally. To do either of these things well, however, requires an automatic component with many of the capabilities of a fully automatic interpretation system [302]. The interactive map interpretation techniques described in Chapter 8, for instance, build upon well-developed automatic interpretation methods. As a result, only a few such systems (e.g. FASTRAK [303]) have reached the market place.

The alternative is to work towards automatic interpretation, but allow for a significant amount of interactive post-processing on the way (e.g. [304]). Most workers in the field acknowledge that fully automatic drawing interpretation is beyond the capability of current methods and create interactive versions of their systems to act as experimental tools during system development and/or to provide consistent semi-automatic post-processing. The attitude to interactivity adopted during the ANON project (Chapter 11) is fairly typical in this regard.

Until fast, highly reliable and generally applicable fully automatic systems can be created, the role of the operator will remain an important issue in the design of any practical drawing interpretation methodology. Current methods can be both useful and impressive, but further research into interactive systems is needed on a number of fronts. First, a deeper understanding is required of exactly what it is that the human operator brings to the task that cannot, or currently is not, embedded in the artificial automatic component. This question is intimately related to the use of domain knowledge: what knowledge, indeed what type(s) of knowledge, do humans have which are not amenable to inclusion in an automatic computer-based drawing interpretation system? This question widens consideration of the role of domain knowledge out to include both the drawing understanding software/hardware and those controlling or working with it. Second, only very little, if anything at all, is currently known about the relationship between features of the interface presented by interactive drawing interpretation systems and the (actual and perceived) usability of those systems. Ergonomic and cognitive ergonomic analysis is routinely applied to other complex interactive computer systems. As far as we are aware neither has ever been used within line drawing interpretation.

Both these issues will become more pressing as the capabilities of drawing interpretation systems extend towards scene formation and the resolution of inconsistency. Present interactive systems support either low-level vectorisation/geometric description or mid-level entity recognition. While human intervention is currently necessary to both these tasks, it seems likely that the true value of a co-operating human user will become more apparent when higher level tasks are considered. It might be argued that, at the time of writing, the operator is often simply used to compensate for the limitations of an under-developed automatic component. Looking to the future, the goal of interactive drawing interpretation systems must surely be to allow user and machine to truly co-operate, with each partner playing to their strengths. The ability of the human visual system to appreciate large-scale, global patterns and visual events quickly may be of much greater value than its skill in deciding which of a set of vectors best extends an existing straight line.

12.4 Performance Measures

It has become accepted in recent years that those active in the field cannot go on developing ever more algorithms for binarising, thinning, vectorising and extracting entities from line drawing images without clear and principled methods of evaluating both the systems they propose and those they would seek to replace. Papers on the subject began to appear in reasonable numbers around 1990; sessions dedicated to performance evaluation have been run at the major conferences in the area since the mid-1990s.

Key features required of any evaluation method are that it should be quantitative, objective and task dependent. This latter request may be unexpected; it might be thought that task-independent evaluation would provide a more absolute measure of a given algorithm's performance. Suppose, however, that the output of a raster to vector conversion system is to be used to support optical character recognition. Different OCR techniques place different requirements on the vectors supplied. The performance of the available vectorisation systems can only reasonably be evaluated against the requirements of a particular OCR scheme [305]. The task at hand effectively provides a context for any evaluation performed. Cordella and Marcelli

[306], for example, evaluate thinning algorithms in the context of shape decomposition.

Liu and Dori [307] view performance evaluation as requiring three elements;

1. ground truth or some way of acquiring ground truth (expected/desired output) for some image set;

2. a sound method of matching actual output to ground truth so that individual constructs may be evaluated;

3. quantitative metrics which measure interesting and indicative attributes of the variation between expected and actual output.

Haralick [308] proposed the first general framework for performance evaluation in image analysis, using thinning as a case study. Many variations on the theme are possible. De Boer and Smeulders [309], for example, discuss the generation, and use in performance evaluation, of artificial drawing images. Hori and Doermann [305] apply an extension of Haralick's methodology to the evaluation of raster to vector conversion methods. More recently, Phillips and Chhabra [310] have proposed an evaluation scheme capable of evaluating a complete raster to vector conversion scheme, not just a single module.

While research in performance evaluation might be concerned to a large extent with the design of metrics tuned to particular problems, the identification of these metrics requires detailed and sometimes deep analysis of the interpretation systems and techniques concerned. Smeulders and de Boer [311], for example, argue that distinctions should be drawn between the method, algorithm and code used to implement a given interpretation stage and that performance analysis should be considered at each stage. This correlates directly with Marr's [32] three-level model comprising computational theory, algorithm and mechanism. By encouraging reflection on both individual systems and the design processes which lead to them, performance analysis may have a greater effect on the long term development of drawing understanding than might be first thought.

12.5 Topics for Future Development

The issues raised in Sections 12.1 to 12.4 above are relevant to line drawing interpretation in the large, seeking to extend system capabilities beyond the entity extraction level or to produce entity recognition tools which are either better designed or perform better than those which have gone before. In beginning our closing chapter with a consideration of such issues we do not mean to imply that further work on the processes underpinning entity recognition is unnecessary. Although the interpretation of line drawings has developed significantly in recent years, many unsolved problems remain. The systems developed to date are generally oriented towards the interpretation of comparatively simple documents. If a given map is even reasonably complicated, for example, interactive operations dominate the interpretation. This is appropriate at the present level of development of the technology, but expensive in both time and money. Hence, for the time being at

Current Issues and Future Developments

least, the development of automatic interpretation techniques suited to various types of maps and engineering drawing will continue.

As we have seen, any successful line drawing interpretation technology must include many processes. The currently available forms of each of these could be improved, in particular by taking into account and/or making better use of the specific properties of the target set of line drawings. In what follows we make some general observations on the present strengths and weaknesses of the techniques employed in each of the major phases of entity recognition and point out current and possible future avenues of development.

At present, the techniques available for scanning, pre-processing and binarising images of line drawings are stable enough for use on good quality line drawings. Improvement is required, however, if they are to be reliably employed in the interpretation of poor quality originals. Any or all of several approaches could be taken, though the common theme here is to make greater use of domain knowledge:

- Improved binarisation of grey level images. Even local, adaptive thresholding does not help when an object's local grey level is greater than the expected background value. More sophisticated binarisation methods, exploiting greater knowledge of line drawings and suitable for application to colour images, need to be developed and integrated into scanning hardware.

- Improved vectorisation of grey level images. By this we mean direct extraction of primitives and entities without a separate binarisation step (Section 7.5, Chapter 11). Grey level thinning is a promising technique needing further research, particularly into ways of dealing effectively with large images. Another promising approach is to employ some kind of parameter space transformation across the whole image in the style of the Hough transform (Section 7.4). Once again, any techniques developed here should ideally be capable of extension to deal with colour images.

- At the system level, only very few knowledge-based drawing interpretation systems exert any top-down control over processes applied prior to vectorisation. There is considerable scope for extending the range of such systems to consider the entire process, from drawing input onwards. In particular, as discussed in Section 12.2, research is needed to determine what kinds and levels of pre-processing are required to best support this kind of knowledge-based drawing interpretation.

Vectorisation is perhaps the most commonly researched component of line drawing interpretation. While many vectorisation methods exist, there remains considerable room for improvement. Existing sequential vectorisation methods are overly sensitive to errors in their input. The development of truly robust methods is a priority. Present vectorisation techniques are also generally considered to be too context dependent. Vector databases are often claimed to provide a generic method of representing line drawings; in practice most vectorisation methods are really only applicable to certain drawing types. This may actually be viewed as either a strength or a weakness. If vector descriptions are all that is required it is clearly a problem.

One might argue, however, that the primary aim of vectorisation is to provide an intermediate data structure that prepares the input drawing for high-level interpretation. Although the goal of producing a universal data structure is a worthy one, it may be that in many cases a higher-level, domain specific representation would be much more useful and actually constitutes the true goal of the endeavour. A vectorisation method tuned to a particular drawing class might therefore be more valuable (Section 12.2).

Existing vectorisation methods generally take a bottom-up approach, starting with pixel analysis and moving towards the extraction of increasingly more complex geometric objects. Once again, there are significant opportunities to develop top-down vectorisation methods which operate within the context of a more global analysis of the drawings and use the results of that analysis to guide vectorisation.

Though it might be said to represent the current state of the art in line drawing interpretation, entity recognition harbours many unresolved issues. Among the more pressing practical goals are:

- robust modelling of complex line objects and their recognition in the presence of other intersecting and overlapping structures;
- robust recognition of symbols, characters, and words with varying fonts, orientations, scale, etc.;
- the recognition and separation from graphics of text blocks on engineering drawings and maps;
- principled methods for the construction and representation of consistent 2D scenes from groups of entities.

Research in the field continues apace and our capabilities in these areas will surely develop in the future. Many prospective solutions to these problems are and will be based on syntactical and hybrid recognition methods. It is clearly important that domain-specific knowledge be applied wherever possible, but this must be done in a principled manner. Lessons may therefore also be learnt from the wider fields of artificial intelligence and knowledge engineering.

Although any given step in the raster to vector conversion and entity extraction processes is liable to improvement, it seems likely that these topics will provide a focus for research in line drawing interpretation for some time yet. A glance at the topics considered at any of the recent conferences in the area supports this suggestion; sessions on pre-processing, vectorisation, entity extraction and systems aimed at specific applications (usually engineering drawings and maps) dominate. Work on conflict resolution, parameterisation and 3D reconstruction is in its infancy by comparison. Significant effort in these directions is, however, required if the full benefits of the work that has gone before are to be gained. It may be that to truly resolve inconsistency it is necessary to recognise domain-specific 2D objects; another area that has received comparatively little attention within the line drawing community (though object recognition has been a major topic in machine vision for nearly 30 years).

Line drawing interpretation has been studied since the very beginning of image processing, analysis and machine vision in the late 1960s and early 1970s. Study has been motivated variously by the need to input specific sets of drawings into specific computer-based tools, interest in drawing interpretation as an instance of pattern recognition or as a forcing domain for research in knowledge-based image understanding. The field is now beginning to show signs of maturity: a sizeable, stable research community; established, dedicated conferences, workshops and journals; some level of consensus in the techniques and approaches adopted; a growing interest in performance evaluation techniques. The last is particularly important. The development of agreed evaluative mechanisms generally marks a shift in a new discipline away from the natural initial exploratory phase and towards a period of consolidation.

Much valuable and interesting work remains to be done. Some of this will be incremental, building upon, for example, previous vectorisation methods to produce new, more accurate and more widely applicable, vector extraction techniques. Some will focus on system level issues like the acquisition, representation and use of domain knowledge or the human-computer interface. Some will grow out of and feed into new developments in pattern recognition. Other projects will draw more heavily on machine vision, computer graphics and artificial intelligence. Whatever the specific developments are, we believe the machine interpretation of line drawing images will be an active and fascinating area of research and system development for many years to come.

References

1. Ebi N., B. Lauterbach and W. Anheier. An image analysis system for automatic data acquisition from coloured maps. Machine Vision and Applications 1994; 7: 148-164.
2. Grimson W. Eric L. Object recognition by computer: the role of geometric constraints, The MIT Press, Cambridge, Massachusetts; London, England, 1990.
3. Fu K.S. Syntactic methods in pattern recognition, Academic Press, 1974.
4. Advanced Imaging Magazine, 1995; 12.
5. Visual Understanding Systems Ltd. Visus VIP. The Netherlands, 1988.
6. Kasturi R. et al. A system for interpretation of line drawings, IEEE Transactions on Pattern Analysis and Machine Intelligence 1990; 12: 978-992.
7. Arias J.F. et al. Interpretation of telephone system manhole drawings. In: Proceedings of the 2nd International Conference on Document Analysis and Recognition, 1993, pp 365-368.
8. Devaux P.M. D.B. Lysak Jr. and R. Kasturi. A Complete System for the Intelligent Interpretation of Engineering Drawings. International Journal on Document Analysis and Recognition 1999; 2: 2.
9. Dori D. Vector-based arc segmentation in the machine drawing understanding system environment. IEEE Transactions on Pattern Analysis and Machine Intelligence 1995; 17: 1057-1068.
10. Liu W. and D. Dori. Genericity in graphics recognition algorithms. In: Proceedings of the 2nd International Workshop, GREC '97, Nancy, France 1997, Lecture notes in Computer Science 1389, Tombre K and A.K. Chhabra (eds) pp 9-20.
11. Liu W. and D. Dori. Stepwise recovery of arc segmentation in complex line environments. International Journal on Document Analysis and Recognition 1998; 1: 1.
12. Initial Graphics Exchange Specification (IGES), US National Bureau of Standards, Geithersburd, MD 20899, Version 3.0, April 1986.
13. Ablameyko S.V. et al. Interpretation of engineering drawings; techniques and experimental results. Pattern Recognition and Image Analysis 1995; 5: 380-401.
14. Joseph S.H. and T.P. Pridmore. Knowledge-directed interpretation of mechanical engineering drawings. IEEE Transactions on Pattern Analysis and Machine Intelligence 1992; 14: 928-940.
15. Vaxiere P. and K. Tombre. Celesstin: CAD conversion of mechanical drawings. IEEE Computer 1992; 25: 46-54.
16. Antoine D., S. Colin and K. Tombre. Analysis of technical documents; the REDRAW system. In: Proceedings of the IAPR Workshop on Structural and Syntactic Pattern Recognition, New Jersey, 1990, pp 192-230.
17. Suzuki S. and T. Yamada. Maris: map recognition input system. Pattern Recognition 1990; 23: 919-933.
18. Lauterbach B., N. Ebi and P. Besslich. PROMAP - a system for analysis of topographic maps. In: Proceedings of the IEEE Workshop on Applications of Computer Vision, 1992, IEEE CS Press, pp 46-55.

19. Alemany J. and R. Kasturi. A computer vision system for interpretation of paper-based maps. In: Proceedings of the SPIE Conference on Applications of Digital Image Processing, 1987, X 829, pp 125-137.
20. Boatto L. et al. An interpretation system for land registry maps. IEEE Computer 1992; 25: 25-33.
21. Ablameyko S.V. et al. System for automatic vectorisation and interpretation of map-drawings. In: Proceedings of the 4th International Conference on Computer Vision, 1993, pp 456-460.
22. Ablameyko S.V., B. Beregov and A. Kryuchkov. Computer-aided cartographic system for map digitising. In: Proceedings of ICDAR 1993, pp 115-118.
23. Shimotsuji S. et al. A robust recognition system for a drawing superimposed on a map. IEEE Computer 1992; 25: 978-992.
24. Oshitani T. and T. Watanabe. Parallel map recognition based on multi-layer partitioned blackboard model. In: Proceedings of the International Conference on Pattern Recognition, 1998, pp 1604-1606.
25. Ah-Soon C. and K. Tombre. Variations of the analysis of architectural drawings. In: Proceedings of the International Conference on Document Analysis and Recognition, 1997, pp 347-351.
26. Hutton G. et al. A strategy for on-line interpretation of sketched engineering drawings. In: Proceedings of the International Conference on Document Analysis and Recognition, 1997, pp 771-775.
27. Kasturi R. and L. O'Gorman. Document image analysis - a bibliography. Machine Vision and Applications 1992; 5: 231-243.
28. Kasturi R., S. Siva, C. Chennubhotla and L. O'Gorman. Document image analysis: an overview of techniques for graphics recognition. In: Pre-Proceedings of the IAPR Workshop on Syntactic and Structural Pattern Recognition, 1990, Murray Hill, pp 192-230.
29. Kasturi R., S. Siva and L. O'Gorman. Techniques for line drawing interpretation: an overview. In: Proceedings of the IAPR Workshop on Machine Vision Applications, 1990, pp 151-160.
30. Arias J.F. and R. Kasturi. Recognition of graphical objects for intelligent interpretation of line drawings. In: Aspects of Visual Form Processing, C. Arcelli, L. Cordella, G. Sanniti di Baja (Eds)., World Scientific 1994, pp 11-32.
31. Brady J.M. Computational approaches to image understanding. ACM Computing Surveys 1982; 14: 3-71.
32. Ballard D.H. and C.M. Brown. Computer Vision. Prentice-Hall, 1982.
33. Marr D. Vision. W.H. Freeman & Co., 1982.
34. Sonka M., V. Hlavac and R. Boyle. Image Processing, Analysis and Machine Vision. Chapman & Hall, 1993.
35. Trucco E. and A. Verri. Introductory Techniques for 3D Computer Vision. Prentice-Hall, 1998.
36. Wallace A.M. Industrial applications of computer vision since 1982. IEE Proceedings 1988; 135: 117-136.
37. Hogg D. Model-based vision: a program to see a walking person. Image and Vision Computing 1983; 1: 2-20.
38. Sanniti di Baja G. Visual Form Representation. In: V. Cantoni (Ed), Human and Machine Vision; Analogies and Divergences, Plenum Press, New York, 1994, pp 115-129.

39. Freeman H. Computer Processing of line drawing images. Computing Surveys 1974; 6: 57-97.
40. Lysak D.B., P.M. Devaux and R Kasturi. View labelling for automated interpretation of engineering drawings. Pattern Recognition 1995; 28: 393-407.
41. Watanabe T. and T. Fukumura. Towards an architectural framework of map recognition. In: Proceedings of the 2nd Asian Conference on Computer Vision, Singapore, 1995, pp 617-621.
42. Hayes-Roth B. A blackboard architecture for control. Artificial Intelligence 1985; 26: 251-321.
43. Tsai R. An efficient and accurate camera calibration technique for 3D machine vision. In: Proceedings of IEEE International Conference on Computer Vision and Pattern Recognition, Florida, USA, 1986, pp 364-374.
44. Jain A. Fundamentals of digital image processing, Prentice-Hall International, 1989.
45. Ablameyko S.V. et al. System for automatic vectorisation and interpretation of graphic images. Pattern Recognition and Image Analysis 1993; 3: 39-52.
46. Rutovitz D. Efficient processing of 2-D images. In: Cantoni V. (ed) Progress in image analysis and processing. World Scientific, Singapore, 1990, pp 229-253.
47. Piper J. Efficient implementation of skeletonisation using interval coding. Pattern Recognition Letters 1985; 3: 389-397.
48. Piper J. Interval skeletons. In: Proceedings of the 11th IAPR International Conference on Pattern Recognition, Hague, 1992, Vol. III, pp 468-471.
49. Advanced Imaging Magazine 1994; 3: 44-47.
50. Lindley C.A. Practical image processing in C. Wiley, New York, 1991.
51. Weska J.S. A survey of threshold selection techniques. Computer Graphics and Image Processing 1978; 7: 259-265.
52. Fu K.S. and J.K. Mui. A survey on image segmentation. Pattern Recognition 1981; 13: 3-16.
53. Sahoo P.K. et al. A survey of thresholding techniques. Computer Vision, Graphics and Image Processing 1988: 41; 233-260.
54. Fernando S.M.X. and D.M. Monro. Variable thresholding applied to angiography. Proceedings of the 6th International Conference on Pattern Recognition, Munich, 1982.
55. Chow C.K. and T. Kaneko. Automatic boundary detection of left ventricles from cineangiograms. Computers in Biomedical Researh 1972; 5: 338-410.
56. Haralick R.M., K. Shammugam and I. Dinstein. Texture features for image classification. IEEE Transactions on Systems, Man and Cybernetics 1973; 3: 610-621.
57. Wang S. and R.M. Haralick. Automatic multithreshold selection. Computer Vision Graphics and Image Processing 1984: 25; 46-67.
58. Doyle W. Operations useful for similarity invariant pattern recognition. Journal of the Association of Computing Machinery 1962; 9: 259-267.
59. Prewitt J.M.S. and M.L. Mendelsohn. The analysis of cell images. Annals of the New York Academy of Science 1966: 128; 1035-53.
60. Rosenfeld A. and A.C. Kak. Digital Picture Processing, Academic Press, New York, 1976.

61. Rosenfeld A. and P. De La Torre. Histogram concavity analysis as an aid to threshold selection. IEEE Transactions on Systems Man and Cybernetics 1983; 13: 231-35.
62. Mason D. et al. Measurement of C-bands in human chromosomes. Computers in Biology and Medicine 1975; 5: 179-201.
63. Otsu N. A threshold selection method from grey level histogram. IEEE Transactions on Systems, Man and Cybernetics 1978; 8: 62-66.
64. Dunn M.E. and S.H. Joseph. Processing poor quality line drawings by local estimation of noise. In: Proceedings of the 4th International Conference on Pattern Recognition 1988, pp 153-162.
65. Ahuja N. and A. Rosenfeld. A note on the use of second order grey level statistics for threshold selection. IEEE Transactions on Systems, Man and Cybernetics 1978; 8: 895-899.
66. Deravi F. and S.K. Pal. Grey level thresholding using second-order statistics. Pattern Recognition Letters 1983; 1: 417-422.
67. Kirby R.L. and A. Rosenfeld. A note on the use of (grey level, local average grey level) space as an aid in threshold selection. IEEE Transactions on Systems, Man and Cybernetics 1979; 9: 860-864.
68. Kittler J. and J. Illingworth. Minimum error thresholding. Pattern Recognition 1986; 19: 41-47.
69. Nagawa Y. and A. Rosenfeld. Some experiments on variable thresholding. Pattern Recognition 1979; 11: 191-204.
70. Fan Jiulin and Xie Winxin. Minimum error thresholding: a note. Pattern Recognition Letters 1997; 18: 705-709.
71. Bartz R.M. The IBM 1975 optical page reader, part 2: video thresholding system. IBM Journal of Research and Development 1968; 12: 354-363.
72. O'Gorman L. and R. Kasturi. Document Image Analysis. IEEE Computer Science Press 1995.
73. Tsai W.H. Moment-preserving thresholding: a new approach. Computer Vision, Graphics and Image Processing 1985; 29: 377-393.
74. Weska J.S. and A. Rosenfeld. Histogram modification for threshold selection. IEEE Transactions on Systems, Man and Cybernetics 1979; 9: 38-52.
75. Watanabe S. et al. An automated apparatus for cancer processing. Computer Vision, Graphics and Image Processing 1974; 3: 350-358.
76. Katz Y.H. Pattern recognition of meteorological satellite cloud photography. In: Proceedings of the 3rd Symposium on Remote Sensing of the Environment, 1965, pp 173-214.
77. Weska J.S., R.N. Nagel and A. Rosenfeld. A threshold selection technique. IEEE Transactions on Computers 1974; C-23: 1322-1326.
78. Wu A., T. Hong and A. Rosenfeld. Threshold selection using quadtrees. IEEE Transactions on Pattern Analysis and Machine Intelligence 1982; 4: 90-94.
79. Yanovwitz S.D and A.M. Bruckstein. A new method for image segmentation. Computer Vision, Graphics and Image Processing 1989; 46: 82-95.
80. Trier O.D. and T. Taxt. Evaluation of binarisation methods for document images. IEEE Transactions on Pattern Analysis and Machine Intelligence 1995; 17: 312-315.
81. Joseph S.H. Processing of line drawings for automatic input to CAD. Pattern Recognition 1989; 22: 1-11.

82. Hori O. and S. Tanigawa. Raster-to-vector conversion by line fitting based on contours and skeletons. In: Proceedings of the 2nd International Conference on Document Analysis and Recognition, Japan, 1993, pp 353-358.
83. Kwok P.C.K. and T. J. Turner. Raster to vector conversion in a map interpretation system. In: Proceedings of IAPR Workshop on Machine Vision Applications, 1990, pp 165-168.
84. Ablameyko S.V. et al. Fast raster-to-vector conversion of large-size 2D line drawings in a restricted computer memory. In: Proceedings of the IAPR Workshop on Machine Vision Applications, Japan, 1992, pp 49-62.
85. Espelid R. A raster-to-vector conversion based on industrial requirements. In: Proceedings of the IAPR Workshop on Computer Vision - Special Hardware and Industrial Applications, Tokyo, 1988, pp 224-228.
86. Kong T.Y. and A. Rosenfeld. Digital topology: introduction and survey. Computer Vision, Graphics, and Image Processing 1989; 48: 353-393.
87. Borgefors G. Applications using distance transforms. In: Arcelli A. et al. (Eds) Aspects of Visual Form Processing. World Scientific, 1994, pp 83-108.
88. Rosenfeld A. Connectivity in digital pictures. Journal of the Association of Computing Machinery 1970; 17: 146-160.
89. Toriwaki J.-I. and S. Yokoi. Distance transformations and skeletons of digitized pictures with applications. In: Kanal L.N. and A.Rosenfeld (eds) Progress in Pattern Recognition, North-Holland Publishing Company, 1981, pp 187-264.
90. Lam L., S.W. Lee and C.Y.Suen. Thinning methodologies - a comprehensive survey. IEEE Transactions on Pattern Analysis and Machine Intelligence 1992; 14: 869-885.
91. Rosenfeld A. and J.L. Pfaltz. Sequential operations in digital picture processing. Journal of the Association of Computing Machinery 1966; 13: 471-494.
92. Borgefors G. Distance transformations in arbitrary dimensions. Computer Vision, Graphics, and Image Processing 1984; 27: 321-345.
93. Borgefors G. Distance transformations in digital images. Computer Vision, Graphics, and Image Processing 1986; 34: 344-371.
94. Danielsson P.E. Euclidean distance mapping. Computer Graphics and Image Processing 1980; 14: 227-248.
95. Dorst L. and P.W. Verbeek. The constrained distance transformation: a pseudo-euclidean, recursive implementation of the lee-algorithm. In: Young I.T. et al. (Eds), Signal Processing III: Theories and Applications, Elsevier Science, 1986, pp 917-920.
96. Arcelli C. and G. Sanniti di Baja. Weighted distance transforms: a characterisation. In: Cantoni V. et al. (Eds) Image Analysis and Processing II, Plenum Press, New York, NY, 1988, pp 205-211.
97. Ragnemalm I. The euclidean distance transform. PhD Thesis, Linkoping University, Sweden, No.304, 1993.
98. Toriwaki J-I., N. Kato and T. Fukumura. Parallel local operations for a new distance transformation of a line pattern and their applications. IEEE Transactions on Systems, Man and Cybernetics 1979; SMC-9: 628-643.
99. Ragnemalm I. and S. Ablameyko. On the distance transform of line patterns. In: Proceedings of the Scandinavian Conference on Image Analysis, 1993, pp 1357-1363.

100. Pridmore T.P. and S.V. Ablameyko. A generalised distance transform for line patterns. Pattern Recognition and Image Analysis 1996; 6: 545-554.
101. Ablameyko S.V., C. Arcelli and G. Sanniti di Baja. Finding and ranking basic structures on complex line patterns. In: D. Dori and A. Bruckstein (Eds),Shape, Structure and Pattern Recognition, World Scientific, Singapore, 1995, pp 33-42.
102. Matheron G. Random sets and integral geometry, John Wiley, 1975.
103. Serra J. Image analysis and mathematical morphology, Academic Press, 1983.
104. Heijmans H. and C. Ronse. The algebraic basis of mathematical morphology I. dilations and erosions. Computer Vision, Graphics, and Image Processing 1990; 50: 245-295.
105. Heijmans H. and C.Ronse. The algebraic basis of mathematical morphology II. openings and closings. Computer Vision, Graphics, and Image Processing - Image Understanding 1994; 54: 74-97.
106. Giardina C. and E. Dougherty. Morphological methods in image and signal processing, Prentice-Hall, 1988.
107. Koskinen L. and J. Astola. Morphological filtering of noisy images. Proceedings of the .SPIE 1990; 1360: 155-165.
108. Serra J. and B. Lay. Algorithms in mathematical morphology, Academic Press, 1988.
109. Serra J. and L. Vincent. An overview of morphological filtering. Circuits, Systems and Signal Processing 1992; 11: 47-108.
110. Dougherty E. Optimal mean-square N-observation digital morphological filters-part I: optimal binary filters. Computer Vision, Graphics and Image Processing - Image Understanding 1992; 55: 36-54.
111. Ablameyko S.V., C. Arcelli and G. Sanniti di Baja. Using distance information for editing binary pictures. In: Proceedings of the 6th Scandinavian Conference on Image Analysis, Finland, 1989, pp 401-407.
112. Black W. et al. A general purpose follower for line structured data. Pattern Recognition 1981; 14: 33-42.
113. Sanniti di Baja G. Well-shaped stable and reversible skeletons from the (3,4)-distance transform. Journal of Visual Communication and Image Representation 1994; 5: 107-115.
114. Arcelli C., L. Cordella and S. Levialdi. Parallel thinning of binary pictures. Electronic Letters 1975; 11: 148-149.
115. Hilditch C.J. Linear skeletons from square cupboards. In: Meltzer B. and D. Michie (Eds), Machine Intelligence 4, American Elsevier, New York, 1969, pp 403-420.
116. Rutovitz D. Pattern recognition. Journal of the Royal Statistical Society 1966; 129: A: 504-530.
117. Stefanelli R. and A. Rosenfeld. Some parallel thinning algorithms for digital pictures. Journal of the Association for Computing Machinery 1971; 18: 255-264.
118. Eckhardt U. and G. Maderlechner. Thinning for document processing. In: 1st International Conference on Document Analysis and Recognition, 1991, pp 490-498.
119. Chen S. and W.H. Hsu. A comparison of some one-pass parallel thinnings. Pattern Recognition Letters 1990; 11: 35-41.

120. Jang B.K. and R.T. Chin. Analysis of thinning algorithms using mathematical morphology. IEEE Transactions on Pattern Analysis and Machine Intelligence 1990; 12: 541-551.
121. Lam L. and C.Y. Suen. An evaluation of parallel thinning algorithms for character recognition. IEEE Transactions on Pattern Analysis and Machine Intelligence 1995; 17: 914-919.
122. Ablameyko S.V. et al. Vectorisation and representation of large-size 2D line drawings. Journal of Visual Communication and Image Representation 1994; 5: 245-254.
123. Arcelli C. and G. Sanniti di Baja. A one-pass two-operation process to detect the skeletal pixels on the 4-distance transform. IEEE Transactions on Pattern Analysis and Machine Intelligence 1989; 11: 411-414.
124. Sanniti di Baja G. and E. Thiel. Skeletonisation algorithm running on path-based distance maps. Image and Vision Computing 1996; 14: 47-57.
125. Arcelli C., L.P. Cordella and S. Levialdi. From local maxima to connected skeletons. IEEE Transactions on Pattern Analysis and Machine Intelligence 1981; 3: 134-143.
126. Arcelli C. and G. Sanniti di Baja. Skeletons of planar patterns. In: Kong T.Y and A. Rosenfeld (Eds), Topological algorithms for digital image processing, North-Holland, 1996, pp 99-143.
127. Arcelli C. and G. Sanniti di Baja. Euclidean skeleton via centre-of-maximal-disc extraction. Image and Vision Computing 1993; 11: 163-173.
128. Russ J.C., The image processing handbook, CRC Press, 1995.
129. Atali D., G. Sanniti di Baja and E. Thiel. Pruning discrete and semicontinuous skeletons. In: C.Braccani, L.De Floriani, G.Vernazza, (Eds), Image Analysis and Processing, Lecture Notes in Computer Science, 974, Springer, 1995, pp 488-493.
130. Arcelli C. and G. Ramella. Finding gray-skeletons by iterated pixel removal. Image and Vision Computing 1995; 13: 159-167.
131. Ramella G. Extracting thin lines in gray-level images by pixel removal and ordered propagation. Pattern Recognition and Image Analysis 1996; 5: 570-576.
132. Arcelli C. and G. Ramella. Sketching a gray-tone pattern from its distance transform. Pattern Recognition 1996; 29: 2033-2045.
133. Abe K., F. Mizutani and C. Wang. Thinning of gray-scale images with combined sequential and parallel conditions for pixel removal. IEEE Transactions on Systems, Man and Cybernetics 1994; 24: 294-299.
134. Yu S.S. and W.H. Tsai. A new thinning algorithm for gray scale images by the relaxation technique. Pattern Recognition 1990; 23: 1067-1076.
135. Tang Y.Y. et al. Extraction of reference lines from documents with gray-level background. In: Proceedings of the 3rd International Conference on Document Analysis and Recognition – ICDAR '95, Vol. 2, 1995, pp 571-574.
136. Wang L. and T. Pavlidis. Detection of curved and straight segments from gray scale topography. CVGIP: Image Understanding 1993; 58: 352-365.
137. Lee S.W. and Y.J. Kim. Direct extraction of topographic features from gray scale character images. In: Proceedings of the 3rd Pacific Rim International Conference on Artificial Intelligence, Vol 2. Beijing, 1994, pp 784-790.

138. Ablameyko S.V., C. Arcelli and G. Ramella. Removing noninformative subsets during vectorization of gray-level line patterns. SPIE Proceedings 1996; 272: 1514-1521.
139. Ramer U. Extraction of line structures from photographs of curved objects. Computer Graphics and Image Processing 1975; 4: 81-103.
140. Kurozumi Y. and W.A. Davis. Polygonal approximation by minimax method. Computer Graphics and Image Processing 1982; 19: 248-264.
141. Ramer U. An iterative procedure for the polygonal approximation of plane curves. Computer Graphics and Image Processing 1972; 1: 244-256.
142. Sklansky J. and V. Gonzalez. Fast polygonal approximation of digitized curves. Pattern Recognition 1980; 12: 327-331.
143. Williams C.M. An efficient algorithm for the piecewise linear approximation of planar curves. Computer Graphics and Image Processing 1978; 8: 286-293.
144. Bimal K.R. and S.R. Kumar. A new approach to polygonal approximation. Pattern Recognition Letters 1991; 12: 229-234.
145. Hemminger T.L. and C.A. Pomalaza-Raez. Polygonal representation: a maximum likelihood approach. Computer Vision, Graphics, and Image Processing 1990; 52: 239-247.
146. Zhang S., L. Li and H. Seah. Vectorization of digital images using algebraic curves. Computing & Graphics 1998; 22: 91-101.
147. Barry P.J. and R.N. Goldman. Interpolation and approximation of curves and surfaces using polya polynomials. CVGIP: Graphical Models and Image Processing 1991; 53: 137-148.
148. Pham B. Quadratic B-splines for automatic curve and surface fitting. Computer Vision, Graphics, and Image Processing 1989; 13: 471-475.
149. Potier C. and C. Vercken. Geometric modeling of digitized curves. In: Proceedings of the 1st International Conference on Document Analysis and Recognition, St Malo, France, 1991, pp 152-160.
150. Foley J.D and A. Van Dam. Fundamentals of Computer Graphics. Addison-Wesley, Reading, Mass., 1982.
151. Pavlidis T. Algorithms for Graphics and Image Processing. Computer Science Press, Rockville, Maryland, 1982.
152. Medioni G. and Y. Yasumoto. Corner detection and curve representation using cubic B-splines. Computer Vision, Graphics, and Image Processing 1987; 29: 267-278.
153. Pridmore T.P., J. Porrill and J.E.W. Mayhew. Segmentation and description of binocularly viewed contours. Image and Vision Computing 1987; 5: 132-138.
154. Porrill J. et al. TINA: a 3D vision system for pick and place. Image and Vision Computing 1988; 6: 91-99.
155. Pavlidis T. A vectorizer and feature extractor for document recognition. Computer Vision, Graphics, and Image Processing 1986; 35: 111-127.
156. Suzuki S. Graph-based vectorization method for line patterns. In: Proceedings of the International Conference on Pattern Recognition, Roma, 1988, pp 616-621.
157. Nagasamy V. and N.A. Langrana. Engineering drawing processing and vectorization system. Computer Vision, Graphics, and Image Processing 1990; 49: 379-397.

References

158. Shih C. and R. Kasturi. Extraction of graphic primitives from images of paper based line drawings. Machine Vision and Applications 1989; 2: 103-113.
159. Hough P.V.C. Methods and means for recognising complex patterns. U.S. Patent 3,069,554, Dec. 18th, 1962.
160. Duda R.O. and P.E. Hart. Use of the hough transformation to detect lines and curves in pictures. Communications of the ACM 1972; 15: 11-15.
161. Kimme C., D. Ballard and J. Sklansky. Finding circles by an array of accumulators. Communications of the ACM 1975; 18: 120-122.
162. Yuen H.K. et al. Comparative study of hough transform methods for circle finding, Image and Vision Computing 1990; 8: 71-77.
163. Tsuji S. and F. Matsumoto. Detection of ellipses by a modified hough transformation. IEEE Transactions on Computers 1978; 27: 777-781.
164. Proctor S. and J. Illingworth. A comparison of the randomised hough transform and a genetic algorithm for ellipse detection. In: E Gelsema and L Kanal (Eds), Pattern Recognition in Practice IV: multiple paradigms, comparative studies and hybrid systems, Elsevier Science Ltd, pp 449-460.
165. Ballard D. Generalising the hough transform to detect arbitrary shapes. Pattern Recognition 1981; 13: 111-122.
166. Canny J. A computational approach to edge detection. IEEE Transactions On Pattern Analysis And Machine Intelligence 1986; 8: 679-698.
167. Lammerts van Beuren G.M. Scannen + vectoriseren of digitaliseren? Een vergelijking. Geodesia 1993; 35: 114-118 (in Dutch).
168. Janssen R.D.T. and A.M. Vossepoel. Adaptive vectorization of line drawing images. Computer Vision and Image Understanding 1997; 65: 38-56.
169. Ablameyko S.V., B. Beregov and A. Kryuchkov. Automatic map digitising: problems and solution. Computing and Control Engineering Journal 1994; 5: 33-39.
170. Van Gool L., P. Dewaele and A. Oosterlinck. Texture analysis anno 1983. Computer Vision, Graphics, and Image Processing 1985; 29: 336-357.
171. Muller M.J. Texture boundaries: important cues for human texture discrimination. In: Proceedings of the IEEE Computer Society Conference on Computer Vision and Pattern Recognition 1986, pp 464-468.
172. Starovoitov V. et al. Binary texture border extraction on line-drawings based on distance transform. Pattern Recognition 1993; 8: 1165-1176.
173. Amin T.J. and R. Kasturi. Map data processing: recognition of lines and symbols. Optical Engineering 1987; 26: 354-358.
174. Morean O.A. and R. Kasturi. Symbol identification in geographical maps. In: Proceedings of the 7th International Conference on Pattern Recognition, 1984, pp 966-967.
175. Bhattacharjee S. and G.Monagan. Recognition of cartographic symbols. In: Proceedings of the IAPR Workshop on Machine Vision Applications, 1994, pp 226-229.
176. Ablameyko S.V. et al. Interactive interpretation of map-drawing images. In: Proceedings of the 2nd IAPR Workshop on Graphics Recognition, Nancy, France, 1997, pp 185-192.
177. Lai C.P. and R. Kasturi. Detection of dashed lines in engineering drawings and maps. In: Proceedings of the 1st International Conference on Document Analysis and Recognition, 1991, pp 507-515.

178. Ohsawa Y. and M. Sakauchi. The automatic recognition system of dotted and broken lines in engineering drawings and maps. In: Proceedings of the International Conference on Electronics, Control and Instrumentation, San Francisco, Calif., 1985, Vol. 2, pp 684-687.
179. Committee on the Exchange of Digital Data (CEDD), International Hydrographic Organization, Format for exchange of digital hydrographic data 1986.
180. IHO DX-90: Specifications for the exchange of digital hydrographic data 1990.
181. ESRI, Understanding GIS: the ARC/INFO method (6th ed.), Redlands, ESRI 1992.
182. Burrough P.A. Principles of geographical information systems for land resources assessment, Clarendon Press, Oxford, 1986.
183. Chrisman N.R., The role of quality information in the long-term functioning of a geographic information system. Cartographica 1984; 21: 79-87.
184. Mead D.A. Assessing data quality in geographic information systems. In: Johannsen C.J. and J.L. Sanders (Eds) Remote sensing for resource management. Soil Conservation Society of America, Ankeny, IA., 1982, pp 51-62.
185. Ablameyko S.V. and G. Aparin. Data validity in map-drawing interpretation. In: Proceedings of the International Conference ICARCV 92, Singapore, 1992, pp CV-17.8.1 - CV-17.8.5.
186. Aparin G. and S. Ablameyko. Quality of recognized maps: how it could be provided? In: Proeedings of the 2nd IAPR Workshop on Graphics Recognition, Nancy, France, 1997, pp 177-184.
187. Elliman D.G. and M. Sen-Gupta. Automatic recognition of linear features, symbols and textured areas within maps. In: Proceedings of the IAPR Workshop on Machine Vision Applications, 1994, pp 239-242.
188. Dori D. et al. Sparse-pixel recognition of primitives in engineering drawings. Machine Vision and Applications 1993; 6: 69-82.
189. Kong B. etal. A benchmark: performance evaluation of dashed-line detection algorithms. In: Proceedings of the International Workshop on Graphics Recognition, Lecture Notes in Computer Science, 1072, 1996, pp 270-285.
190. Dori D., L. Wenyin and M. Peleg. How to win a dashed line detection contest. In: Proceedings of the International Workshop on Graphics Recognition, Lecture Notes in Computer Science, 1072, 1996, pp 286-300.
191. Liu W.Y. and D. Dori. A protocol for performance evaluation of line detection algorithms. Machine Vision and Applications 1997; 9: 240-250.
192. Phillips I.T., J. Liang and R. Haralick. A performance evaluation protocol for engineering drawing recognition systems. In: Proceedings of the 2nd International Workshop on Graphics Recognition, Nancy, France, 1997, pp 333-346.
193. Semenkov O. et al. Information processing and display in raster graphical systems, Nauka i Tekhnika, Minsk, 1989.
194. Ilg M. Knowledge-based understanding of roadmaps and other line images. In: Proceedings of the 10th International Conference on Pattern Recognition, 1990, pp 282-284.

195. Ilg M. and R. Ogniewicz. Knowledge-based interpretation of roadmaps based on symmetrical skeletons. In: Proceedings of the IAPR Workshop on Machine Vision Applications, Japan, 1990, pp 161-164.
196. Hayakawa T. et al. Recognition of roads in an urban map by using the topological road-network. In: Proceedings of the IAPR Workshop on Machine Vision Applications, Japan, 1990, pp 215-218.
197. Musavi M.T. et al. A vision based method to automate map processing. Pattern Recognition 1988; 21: 319-326.
198. Deseilligny M.P., H. Le Men and G. Stamon. Map understanding for GIS data capture: algorithms for road network graph reconstruction. In: Proceedings of the 2nd International Conference on Document Analysis and Recognition, Japan, 1993, pp 676-679.
199. Nagao A., T. Agui and M. Nakajima. An automatic road vector extraction method from maps. In: Proceedings of the 9th International Conference on Pattern Recognition, Italy, 1988, pp 585-587.
200. Ablameyko S.V. et al. Knowledge-based interpretation of roads and correlated objects on map-drawings. In: Proceedings of ICARCV '94 the International Conference, Singapore, 1994, pp 343-347.
201. Shipley T. and M. Shore. The human texture visual field: fovea to periphery. Pattern Recognition 1990; 23: 1215-1221.
202. Unser M. and M. Eden. Multiresolution feature extraction and selection for texture segmentation. IEEE Transactions on Pattern Analysis and Machine Intelligence 1989; 11: 717-728.
203. Reed T.R., H. Wechsler and M. Werman. Texture segmentation using a diffusion region growing technique. Pattern Recognition 1990; 23: 953-960.
204. Kashyap R.L. and K.B. Eom. Texture boundary detection based on the long correlation model. IEEE Transactions on Pattern Analysis and Machine Intelligence 1989; 11: 58-67.
205. Govindan V.K. and A.P. Shivaprasad. Character recognition - a review. Pattern Recognition 1990; 23: 671-683.
206. Yamamoto K., H. Yamada and S. Muraki. Symbol recognition and surface reconstruction from topographic maps by parallel method. In: Proceedings of the 2nd International Conference on Document Analysis and Recognition, 1993, pp 914-917.
207. Shiku O. et al. Extraction of slant character candidates from maps using circular templates. In: Proceedings of the 3rd International Conference on Document Analysis and Recognition, 1995: pp 936-939.
208. Fletcher L.A. and R. Kasturi. A robust algorithm for text separation from mixed text/graphics images. IEEE Transactions on Pattern Analysis and Machine Intelligence 1988; 10: 910-918.
209. Nakamura A. et al. A method for recognizing character strings from maps using linguistic knowledge. In: Proceedings of the 2nd International Conference on Document Analysis and Recognition, 1993, pp 561-564.
210. Pierrot-Deseilligny M., H. Le Men and G. Stamon. Character-string recognition on maps: a method for high-level recognition. In: Proceedings of the 3rd International Conference on Document Analysis and Recognition, 1995, pp 249-252.

211. Luo H., G. Agam and I. Dinstein. Directional mathematical morphology approach for line thinning and extraction of character strings from maps and line drawings. In: Proceedings of the 3rd International Conference on Document Analysis and Recognition, 1995, pp 257-260.
212. Anegawa M. et al. A system for recognizing numeric strings from topographical maps. In: Proceedings of the 3rd International Conference on Document Analysis and Recognition, 1995, pp 940-943.
213. Samet H. and A. Soffer. A legend-driven geographic symbol recognition system. In: Proceedings of the International Conference on Pattern Recognition, 1994, pp 350-354.
214. Chen N.A., N. Langrana and A.K. Das. Perfecting vectorized mechanical drawings. Computer Vision and Image Understanding 1996; 63: 273-286.
215. Filipski A. J. and R. Flandrena. Automated conversion of engineering drawings to CAD form. Proc IEEE 1992; 80: 1195-1209.
216. Kung L.-S. and J.-C. Samin. A procedure of reading mechanical engineering drawings for CAD applications. Signal Processing 1993; 32: 191-200.
217. Yu Y.H. A. Samal and S.C. Seth. A system for recognizing a large class of engineering drawings. IEEE Transactions on Pattern Analysis and Machine Intelligence 1997; 19: 868-890.
218. Langrana N.A., Y.A. Chen and A.K. Das. Feature identification from vectorized mechanical drawings. Computer Vision and Image Understanding 1997; 68: 127-145.
219. Lu Z.Y. Detection of text regions from digital engineering drawings. IEEE Transactions on Pattern Analysis and Machine Intelligence 1998; 20: 431-439.
220. Messmer B.T. and H. Bunke. Automatic learning and recognition of graphical symbols in engineering drawings. In: Proceedings of the International Workshop on Graphics Recognition, Lecture Notes in Computer Science, 1996, Vol. 1072, pp 123-134.
221. Gao J. et al. Segmentation and recognition of text in engineering drawings. In: Proceedings of the 3rd International Conference on Document Analysis and Recognition, 1995, Vol. 1, pp 528-531.
222. Dori D. and Y. Velkovitch. Segmentation and recognition of dimensioning text from engineering drawings. Computer Vision and Image Understanding 1998; 69: 196-201.
223. Kultanen P., E. Oja and L. Xuf. Randomized Hough transform (RHT) in engineering drawing vectorization system. In: IAPR Workshop on Machine Vision Applications, 1990, pp 173-176.
224. Rosin P.L. and G.W. West. Segmentation of edges into lines and arcs. Image and Vision Computing 1989; 7: 109-114.
225. Arias J.F. and R. Kasturi. Efficient extraction of primitives from line drawings composed of horizontal and vertical lines. Machine Vision and Applications, 1997; 10: 214-221.
226. Bixler J.P., L.T. Watson and J.P. Sanford. Spline-based recognition of straight lines and curves in engineering line drawings. Image and Vision Computing 1988; 6: 262-269.
227. Antoni D. CIPLAN: A model-based system with original features for understanding French plats. In: Proceedings of the International Conference on Document Analysis and Recognition, 1991, Vol. 2, pp 647-655.

228. Min W., Z. Tang and L. Tang. Recognition of dimensions in engineering drawings based on arrowhead-match. In: Proceedings of the 2nd International Conference on Document Analysis and Recognition, 1993, pp 373-376.
229. Lin C. and C.K. Ting. A new approach for detection of dimensions set in mechanical drawings. Pattern Recognition Letters 1997; 18: 367-373.
230. Das A.K. and N.A. Langrana. Recognition of dimension sets and integration with vectorized engineering drawings. In: Proceedings of the 3rd International Conference on Document Analysis and Recognition, 1995, Vol. 1, pp 347-350.
231. Lai C.P. and R. Kasturi. Detection of dimension sets in engineering drawings. IEEE Transactions on Pattern Analysis and Machine Intelligence 1994; 16: 848-855.
232. Dori D. A syntactic/geometric approach to recognition of dimensions in engineering machine drawings. Computer Vision, Graphics, and Image Processing 1989; 47: 271-291.
233. Dori D. Syntax-enhanced parameter learning for recognition of dimensions in engineering machine drawings. International Journal of Robotics and Automation 1990; 5: 59-67.
234. Min W., Z. Tang and L. Tang. Using web grammar to recognize dimensions in engineering drawing. Pattern Recognition 1993; 26: 1407-1416.
235. Collin S. and D. Colnet. Syntactic analysis of technical drawing dimensions. International Journal of Pattern Recognition and Artificial Intelligence 1994; 8: 1131-1148.
236. Colin S. and P. Vaxiviere. Recognition and use of dimensioning in digitized industrial drawing. In: Proceedings of the 1st International Conference on Document Analysis and Recognition, 1991, pp 161-169.
237. Collin S. and D. Colnet. Analysis of dimension in mechanical engineering drawings. In: Proceedings of the IAPR Workshop on Machine Vision Applications, 1990, pp 105-108.
238. Dori D. and A. Pnueli. The grammar of dimensions in machine drawings. Computer Vision, Graphics, and Image Processing 1988; 42: 1-18.
239. Capelades M.A. and O.I. Camps O.I. Functional parts detection in engineering drawings: looking for the screws. In: Proceedings of the International Workshop on Graphics Recognition, Lecture Notes in Computer Science, 1996, Vol. 1072, pp 246-259.
240. Vaxiviere P. and K. Tombre. Interpretation of mechanical engineering drawing for paper-CAD conversion. In: Proceedings of the IAPR Workshop on Machine Vision Applications, 1990, pp 203-206.
241. Joseph S.H., T.P. Pridmore and M.E. Dunn. Grammar-driven interpretation of engineering drawings. In: Proceedings of the 4th Alvey Vision Conference, University of Manchester, 1988, pp 237-242.
242. Pridmore T.P. and S.H. Joseph. Using schemata to interpret images of mechanical engineering drawings. In: Proceedings of the 9th European Conference on Artificial Intelligence, Stockholm, 1990, pp 515-521.
243. Joseph S.H. and T.P. Pridmore. A system for the interpretation of images of graphics. In: Proceedings of the 1990 IAPR Workshop on Syntactic and Structural Pattern Recognition, New Jersey, 1990.

244. Pridmore T.P. and S.H. Joseph. Integrating visual search with visual memory in a knowledge-directed image interpretation system. In: Proceedings of BMVC90, Oxford, 1990, pp 367-373.
245. Joseph S.H. Segmentation and aggregation of text from images of mixed text and graphics. In: Klette R. (Ed), Research in Informatics vol. 5, 1991, pp 265-271.
246. Joseph S.H., T.P. Pridmore and M.E. Dunn. Towards the automatic interpretation of mechanical engineering drawings. In: Bartlett A. (Ed), Computer Vision and Image Processing, Kogan Page, 1989.
247. Binford T.O. A survey of model-based image analysis systems. Int J. Robotics Research 1982; 1: 18-63.
248. Mulder J.A., A.K. Mackworth and W.S. Havens. Knowledge structuring and constraint satisfaction: the MAPSEE approach. IEEE Transactions on Pattern Analysis and Machine Intelligence 1988; 10: 866-879.
249. Nagao M., T. Matsuyama and H. Mori. Structural analysis of complex aerial photographs. In: Proceedings of the 6th IJCAI, 1979, pp 610-616.
250. Mulder J.A. Discrimination vision. Computer Vision, Graphics and Image Processing 1988; 43: 313-336.
251. McKeown D.M., W.A. Harvey and J. McDermott. Rule-based interpretation of aerial imagery. IEEE Transactions on Pattern Analysis and Machine Intelligence 1985; 7: 570-585.
252. Sakai T., T. Kanade and Y. Ohta. Model-based interpretation of outdoor scenes. In: Proceedings of the 3rd IJCPR, 1976, pp 581-585.
253. Matsuyama T. and V. Hwang. SIGMA: a framework for image understanding - integration of bottom-up and top-down analyses. In: Proceedings of the 9th IJCAI, 1985, pp 908-915.
254. Kohl C.A., A.R. Hanson and E.M. Riseman. A goal-directed intermediate level executive for image interpretation. In: Proceedings of the 10th IJCAI, 1987, pp 811-814.
255. Matsuyama T. and T.S-S. Hwang. SIGMA, a knowledge-based aerial image understanding system, Plenum Press, 1990.
256. Mackworth A.K. Vision research strategy: black magic, metaphors, mechanisms, miniworlds and maps. In Hanson A.R. and E. Riseman (Eds), Computer Vision Systems, Academic Press, 1978.
257. Hanson A.R. and E.M. Riseman. VISIONS: a computer system for interpreting scenes. In Hanson A.R. and E. Riseman (Eds), Computer Vision Systems, Academic Press, 1978.
258. Elliman D.G. and I.T. Lancaster. A review of segmentation and contextual analysis techniques required for automatic text recognition. Pattern Recognition 1990; 23: 337-346.
259. Cheetham S.J. The automatic extraction and classification of curves from conventional line drawings. PhD Thesis. University of Sheffield. 1988.
260. Yachida M., M. Ikeda and S. Tsuji. A knowledge-directed line finder for analysis of complex scenes. In: Proceedings of the 6th IJCAI, 1979, pp 984-991.
261. Neisser U. Cognition and Reality: Principles and Implications of Cognitive Psychology. W.H. Freeman. 1976.
262. Johnson S.C. Yacc - yet another compiler compiler. Computer Science Technical Report No. 32, Bell Laboratories, Murray Hill, New Jersey, 1975.

263. Henderson T.C. and A. Samal. Shape grammar compilers. Pattern Recognition 1986; 19: 279-288.
264. Rao A.R. and A.K. Jain. Knowledge representation and control in computer vision systems. IEEE Expert 1988; Spring: 64-79.
265. McKeown D.M., W.A. Harvey and L.E. Wilson. Automating Knowledge Acquisition for Aerial Image Interpretation. Computer Vision, Graphics and I Image Processing 1989; 46: 37-81.
266. Brooks R.A. Symbolic reasoning among 3D models and 2D images. Artificial Intelligence 1981; 17: 285-348.
267. Kernighan B.W. and D.M. Ritchie. The C Programming Language, Prentice-Hall, 1978.
268. McDermott J. Making Expert Systems Explicit. In: Kugler H.J. (Ed), Information Processing '86, Elsevier Science Publishers (North-Holland), 1986.
269. Tropf H. and I. Walter. An ATN model for recognition of solids in single images. In: Proceedings of the 8th IJCAI, 1983, pp 1094-1097.
270. Tropf H. Analysis-by-synthesis search for semantic segmentation applied to workpiece recognition. In: Proceedings of the 5th ICPR, 1980, pp 241-244.
271. Hattich W. Recognition of overlapping workpieces by model directed construction of object contours. Digital Systems for Industrial Automation 1982; 1: 223-239.
272. Glicksman J. Using multiple information sources in a computational vision system. In: Proceedings of the 8th IJCAI, 1983, pp 1078-1080.
273. Nazif A.M. and M.D. Levine. Low-level image segmentation: an expert system. IEEE Transactions on Pattern Analysis and Machine Intelligence 1984; 6: 555-577.
274. Wu W. and M. Sakauchi. A multipurpose drawing understanding system with flexible object-oriented framework. In: Proceedings of the 2nd International Conference on Document Analysis and Recognition, Japan, 1993, pp 870-873.
275. Doyle J. A truth maintenance system. Artificial Intelligence 1979; 12: 231-272.
276. Draper B.A. et al. The schema system. International Journal of Computer Vision 1989; 2: 209-250.
277. Aggazani D. et al. Control strategies for error handling in a knowledge-based system for image processing and recognition. SPIE Applications of Artificial Intelligence 1990; 1293; 31-42.
278. Dellepiane S., G. Venturi and G. Vernazza. Model generation and model matching of real images by a fuzzy approach. Pattern Recognition 1992; 25: 115-137.
279. Roller D., F. Schoner and A. Verroust. Dimension-driven geometry in CAD: a survey. Theory and Practice of Geometric Modelling 1989; 509-523.
280. Ablameyko S.V. and A.G. Gorelik. Parameterisation of automatically vectorised engineering drawings. In: Proceedings of the 6th International Conference on Advanced Computer Systems, Szczecin, Poland, 1999, pp 32-36.
281. Gorelik A.G. Graphical system based on the relations. Proceedings of the National Academy of Sciences of Belarus, physics - mathematics sciences 1996; 2: 105-111.
282. Iwata K., M. Yamamoto and M. Iwasaki. Recognition system for three-view engineering drawings. Lecture Notes in Computer Science 1988; 31: 240-249.

283. Kuo M.H. Reconstruction of quadric surface solids from three-view engineering drawings. Computer-Aided Design 1998; 30: 517-527.
284. Dori D. and K. Tombre. From engineering drawings to 3-D CAD models: are we ready now? Computer-Aided Design 1995; 29: 243-254.
285. Tombre K. Graphics recognition - general context and challenges. Pattern Recognition Letters 1995; 16: 883-891.
286. Wesley M.A. and G. Markovsky. Fleshing out projections. IBM Journal of Research and Development 1981; 25: 934-954.
287. Ah-Soon C. and K.Tombre. A Step towards 3-D CAD models from engineering drawings. In: Proceedings of the 3rd International Conference on Document Analysis and Recognition, Montreal, Canada, 1995, pp 331-334.
288. Iwata K., M. Yamamoto and M. Iwasaki. Recognition system for three-view engineering drawings. Lecture Notes in Computer Science 1988; 31: 240-249.
289. Kim C. et al. Understanding three-view drawings of mechanical parts with curved shapes. In: Proceedings of the IEEE International Conference on Systems Engineering, Kobe, Japan, 1992, pp 238-241.
290. Chen Z. and D.-B. Perng. Automatic reconstruction of 3D solid objects from 2D orthographic views. Pattern Recognition 1988; 21: 439-449.
291. Yan Q., C. L. P. Chen and Z. Tang. Efficient algorithm for the reconstruction of 3D objects from orthographic projections. Computer-Aided Design 1994; 26: 699-717.
292. Gloger J.M. et al. An AL-aided system for the conversion of paper-based technical drawings into a CAD format. In: Proceedings of the IAPR Workshop on Machine Vision Applications, Tokyo, Japan, 1992, pp. 299-302.
293. Kitajima K. and M. Yoshida. Reconstruction of CSG solid from a set of orthographic three views. In: Proceedings of the IEEE International Conference on Systems Engineering, Kobe, Japan, 1992, pp 220-224.
294. Marti E. et al. A system for interpretation of hand line drawings as three-dimensional scene for CAD input. In: Proceedings of the 1st International Conference on Document Analysis, Saint-Malo, France, 1991, pp 472-480.
295. Tomiyama K. and K. Nakaniwa. Recognition of 3D solid model from three orthographic views - top-down approach. In: Proceedings of the International Workshop on Graphics Recognition 1996, Lecture Notes in Computer Science, Vol. 1072, pp 260-269.
296. Ablameyko S.V. et al. 3D object reconstruction from engineering drawing projections. Computing and Control Engineering Journal 1999; 10: 277-284.
297. Den Hartog J. A framework for knowledge-based map interpretation. PhD thesis, Technische Universiteit Delft, September 1995.
298. Joseph S.H., S. Ablameyko and T.P. Pridmore. Knowledge-based interpretation of engineering drawings and maps. In: Proceedings of the International Workshop on Graphics Recognition, USA, 1995, pp 189-199.
299. Ablameyko S.V. and O. Frantskevich. Knowledge-based technique for map-drawing interpretation. In: Proceedings of the 4th International Conference on Image Processing and its Applications, Maastricht, The Netherlands, 1992, pp 550-554.
300. Ablameyko S.V., G. Aparin, and O. Frantskevich. Knowledge representation and use in map interpretation systems. In: Proceedings of the European Robotics and Intelligent Systems Conference, Malaga, Spain, 1994, pp 297-307.

References

301. Hofer-Alfeis J. Conversion of existing mechanical drawings to CAD data structures: state of the art. In: Proceedings of CAPE 86, Copenghagen, 1986.
302. Hutton G.M. et al. Using temporal information to improve the interpretation of sketched engineering drawings in the Designer's Apprentice. In: Proceedings of the 2nd IAPR Workshop on Graphics Recognition, Nancy, France, 1997, pp 103-110.
303. Fulford M.C. The FASTRAK automatic digitising system. Pattern Recognition 1981; 14: 65-74.
304. Goodson K.J. and P.H. Lewis. A knowledge-based line recognition system. Pattern Recognition Letters 1990; 11: 295-304.
305. Hori O. and D.S. Doermann. Quantitative measurement of the performance of raster to vector conversion algorithms. In: Proceedings of the International Workshop on Graphics Recognition, Penn State University, 1995, pp 272-281.
306. Cordella L.P. and A. Marcelli. An alternate approach to the performance evaluation of thinning algorithms for document processing applications. In: Proceedings of International Workshop on Graphics Recognition, Penn State University, 1995, pp 282-290.
307. Liu W. and D. Dori. A proposed scheme for performance evaluation of grahics/text separation algorithms. In: Proceedings of the 2nd International Workshop on Graphics Recognition, Nancy, France, Lecture Notes in Computer Science 1389, 1987, pp 359 – 371.
308. Haralick R.M. Performance characterisation in image analysis - thinning, a case in point. Pattern Recognition Letters 1992; 13: 5-12.
309. De Boer C. and A.W.M. Smeulders. Simulating graphics for performance analysis purposes by combining real world background and synthetic foreground data. In: Proceedings of the 2nd International Workshop on Graphics Recognition, Nancy, France, Lecture Notes in Computer Science 1389, 1987, pp 347- 358.
310. Phillips I.T. and A.K. Chhabra. Empirical performance evaluation of graphics recognition systems. IEEE Transactions on Pattern Analysis and Machine Intelligence 1999; 21: 849-869.
311. Smeulders A.W.M. and C. de Boer. Design and performance in object recognition. In: Proceedings of the 2nd International Workshop on Graphics Recognition, Nancy, France, Lecture Notes in Computer Science 1389, 1987, pp 335-346.

Index

3D reconstruction 39, 75, 250, 253, 260
3D shape 14, 25, 27, 37
4-connectivity 77-9, 107
8-connectivity 77-9, 87, 92, 103, 107-8, 111, 117, 128

A priori knowledge 9, 14, 17-8, 23, 39-40, 62, 98, 175, 196, 200, 203, 209, 212, 253-4, 257-61
Adaptive threshold selection 59
ANON 17, 89, 133, 164, 209-56
ARC/INFO 159, 272
Architectural drawings 4, 18
Arcs 16-8, 26, 34, 124, 128, 185, 191-3, 200, 246, 255, 274
Area objects 143-7, 180
Arrowheads 16-7, 26, 34, 42, 62, 102, 191-2, 199-200, 210, 225, 228, 234
Auxiliary information 53, 145-6, 256

Bimodal intensity distributions 61, 66, 71-3
Binarisation 57-9, 61, 64, 71-5, 80, 101, 117, 131, 137, 161, 214, 253, 259, 266
Black layer 145, 169-73
Blackboard 5, 18, 42-3, 225, 253, 264-5
Blind vectorisation 141
Blocks, see Text blocks
Bottom-up processing 17, 39-43, 133, 169, 186, 210-12, 222, 225, 235, 260, 276
Boundary approximation 53
Bounding contours 12, 36, 75, 146-7
Branch points 80
Bridges 37
Buildings 4, 11, 36, 169-70, 174, 179, 195, 244

CAD/CAM 1, 4, 16, 26
Cadastral maps 8-11, 157
Camera calibration 45, 265

Cartographic objects 14, 18, 24-6, 36, 75, 143-51
CCD cameras 45-47
CELESTIN 195
Chain code 52-53
Chained lines 133, 155, 214-19, 222-4, 238
Chamfer distance 82-3, 114
Chessboard distance 78-82, 114
CIPLAN 195, 274
Circles 16-7, 26, 34-5, 119, 133, 147, 187, 192-5, 214, 229, 247, 271
Circular arcs, see Arcs
City block distance 78-9
City maps 11
Closed contours 17, 164, 192, 251
Closing, morphological 91-3, 96
Cognitive ergonomic analysis 257
Coincidence link 236-8
Complex graphical primitives 34-7, 189-91
Computer-supported co-operative work (CSCW) 5
Computer vision 13-5, 19, 30, 38, 45, 209-12, 217, 225, 240, 244, 253, 264, 276-7
Conflict resolution 43, 242-4, 256, 260
Conics 128
Connected components 30, 51, 79-80, 97-8, 101, 137-8, 146, 180, 189-91, 195-6
Connected objects 146
Connectivity number 80
Connectivity point 80
Constrained distance transform 83-4
Contour maps 37
Contour pixels 77, 95, 98, 102-3, 109-10, 117-9
Contouring 75-6, 101-6, 121, 140, 170
Co-operative vectorisation 142
Crosshatched boundaries 198
Crosshatching 9, 17, 33-5, 148, 195-7, 207, 212

Crossing number 79-80, 127
Curvature 33, 89, 124-7, 191, 201
Curvature extrema 125-7
Cycle of perception 211, 215-6, 240

Dashed lines 16, 35, 126, 152, 163, 192, 226-9, 271
Data quality 160-1, 272
Defect reduction 140-1
Dilation, morphological 90-1, 94, 177
Dimensioned objects 35, 238
Dimensioning 16-7, 26, 34-5, 97-8, 105, 120, 182, 187, 192, 199-200, 203-7 212, 218, 224-33, 237-40, 243, 246-9, 253, 267, 274-5
Direct vectorisation 74, 133, 214
Distance map 81-6, 95-6, 114, 177-8
Distance transform 81-5, 88, 93-5, 109, 113, 116, 177-80, 267-71
Distance transform for line patterns (DTLP) 83-6, 116
Document image statistics 61
Document scanners 6, 27, 46-7, 64
Domain knowledge, see A priori knowledge
Drawing entities 14, 18, 26-7, 33-5, 73-5, 143, 152, 188, 212, 224, 228, 231, 243, 254
Drawing memory 233-42
Drum scanners 47
DTLP, see Distance transform for line patterns
DXF 16, 31, 159, 192, 247-8
Dyeline copies 58, 67
Dynamic adaptive thresholding 60
Dynamic image partitioning 29, 104

Electrical CAD 4
Electrical schematics 3
Ellipses 8, 26, 133, 271
End points 28-32, 80, 85, 89, 115-6, 119, 128, 138-41, 166-8, 173, 181-2, 187, 193, 245
Entity extraction 26-7, 31, 39, 74-5, 161, 212, 225, 243-4, 253-60
Erosion, morphological 90-1, 94, 110-1, 115, 191, 199

Euclidean distance 77-8, 81-2, 182, 267
Expert Systems, see Knowledge-based systems

Feature-based segmentation 124
Flatbed scanners 47
Flow charts 3
Follow-glass 152
Forests, 143, 147-8, 179-80
Frames 16-7, 212
see also Schemata

Gaussian weighted averaging 125
GDTLP, see Generalised distance transform for line patterns
Generalised distance map for line patterns 86
Generalised distance transform for line patterns (GDTLP) 86-9
Geographic information systems (GIS) 4, 8, 15, 24-6, 34-7, 159-61, 272-3
Geographical survey maps 11
GIF format images 53-54
GIS, see Geographic information systems
Global thresholding 57
Global vectorisation 121, 131
GOLDIE 211, 241
Graphical input systems 9
Grey level histograms 60-74, 266
Grey level skeletonisation 117
Grey level statistics 58, 266

Half-width 66-7
Heads-up vectorisation 142
Hi-fi taxonomy 55
Hilditch algorithm 111
Histogram modification 72
Hough transform 130-3, 163, 192, 259, 274
Human vision 13

IGES 16, 192, 263
Image compression 29, 50
Image enhancement 50
Image file formats 50, 54-5

Image partitioning 28, 73
Image understanding, see Computer vision
Interactive map interpretation 142-5, 152-4, 157, 162, 256
Interior pixels 77, 103
Internal number 105
Isolated point 80, 128-9
Isolines 12, 18, 36, 143-7, 152, 163-8, 254
Iterative end-point fit 122-5
Iterative line approximation 122

KFill algorithm 98-9
Knowledge-based systems 4, 14-15, 42
Knowledge-based vision 14, 209

Land register maps 18, 196
Leader lines 34, 225, 231, 234, 237, 243
Line following 10-4, 109-10, 121-3, 127-8
Line objects 105, 143-8, 152, 157-8, 164-5, 169, 260
Line textures 147, 150
Line width histogram 189, 199
Local thresholding 58, 74
Local vectorisation 121, 128
Logical filters 98

Macrotexture 149, 176
Manual digitisation 6-8, 26, 144-6, 152, 157, 256
Map layers 11, 17, 145, 169
MARIS 17
Mathematical morphology 76, 90, 268-69, 274
Maximal discs 101, 108, 113-5
MDUS 16
Media types 49
Medial axes 11, 101-2, 113-5, 169-70
Medial axis transform 101-2, 113-5, 169
Microtexture 150
Minimax algorithms 123-4
Minimum error thresholding 64-6
Minkowski addition 90

Minkowski subtraction 90
Mixed connectivity 79-80, 108
Mode method 61
Model-based vision 14, 264
Modified distance transform for line patterns 85, 89
Modified run-length code 51
Multi-level thresholding 60

Net textures 150, 179
Node points 18, 119, 122, 191
Noise reduction 16, 50, 74-6, 91-6, 115, 125, 163, 189
Non-maximal suppression 125-7

Opening, morphological 90-96
Optical character recognition 49, 257
Overhead scanners 47-8

Parallel lines 9, 17, 33, 150, 169, 192, 195, 231
Parallel thinning 110
Parameter space 131-3, 259
Parameterisation 41, 130-2, 186, 245-8, 253, 260
Pattern recognition 15, 26, 51, 130, 164, 175, 188, 217, 240, 244, 254, 261-5
Performance evaluation 164, 257-8, 261, 272, 279
Perpendicular-bisector tracing algorithm 192
Physical outlines 33, 200, 223, 226-33, 238
Piece-part drawings 4, 42, 74, 209, 241
Point texture 150
Point-dependent thresholding 60, 64, 68
Principle of least commitment 74, 133
Projections 38-9, 246, 250-2, 278
PROMAP 17, 263
Pruning 116, 269
P-tile method 61, 69

Quadtrees 51-3, 72, 266

Raster images 7-8, 24, 27-9, 159, 196, 199
Raster to vector transformation 26, 75-6, 102, 117-9, 143, 161, 170, 188
REDRAW 17, 195, 263
Region dependent thresholding 60-2, 71-3
Relative entropy 66
Relief maps 37
Reverse distance transform 95, 178
Rivers 36-7, 150, 179, 244
Road layer 145, 170-5
Roads 12, 17-8, 143, 146, 152, 169-75, 179, 187, 244, 254-5, 273
Roadside objects 173-5
Run-length encoding 51, 53-4, 104, 111

Scan line methods 102-4, 109-11, 128
Scanner resolution 49
Scanning speed 50
Scatter plots 62, 71
Scenes 9-10, 24-6, 37, 40, 61, 143-5, 244, 260, 276
Schema fusion 239
Schemata 17, 212-44, 275
  see also Frames
Sectioning lines 16
Self-supervised arrowhead recognition 199
Semantic nets 233, 240
Sequential thinning 109
Sheetfed scanners 47-9
SIGMA 211, 241, 244, 276
Skeletonisation 33, 75-6, 91, 101, 109, 115-8, 121, 140, 170-2, 214, 265
Smoothing 59, 66, 98, 125
SPAM 211, 218, 241, 244
Spline approximation 124, 270
Standards 11, 33, 53, 56, 159, 211
State variable 214-15, 223-4
Static adaptive thresholding 60-1
Static image partitioning 29
Structural pattern recognition 164, 181, 188

Structural textures 149
Structuring element 90-1, 94-6, 191
Symmetry 35, 185-8, 200, 205

Templates 109, 153, 165-6, 273
Text blocks 5, 17, 23, 34, 51, 201-7, 212, 233, 260
Text recognition 5, 180, 185, 276
Text/graphics separation 16
Texture 10, 14, 37, 147-50, 176-80, 271-3
Texture boundaries 176-9
Textured objects 148, 176-8
Thinning 32, 42, 84, 101-2, 108-12, 115-18, 122, 128, 131, 210, 257-9, 268-9, 274, 279
Thresholding 25-7, 32, 39, 42, 50, 57-74, 95, 116-7, 127, 133-5, 176, 189, 216, 220, 259, 265-6
TIFF format images 16, 53-5
Top-down processing 17, 39, 43, 122, 133-5, 169, 212-4, 222, 225, 235, 255, 259-60, 276-8
Topographic maps 11, 18, 157-9, 166, 263, 273
Tracking 28, 42, 102-6, 133-5, 152, 166-72, 203, 210, 214-5, 218-26, 229, 241

Unconnected objects 146
Universal entity representations 17, 34-7
Urban maps 169

Vector representations 8, 17-8, 22, 33, 75-6, 102, 115, 129, 137, 140, 152, 165, 193-6, 205
Vectorisation 9, 15-8, 21-7, 30, 39-42, 51-3, 57, 74-6, 102, 115, 119-24, 127-30, 133-5, 140, 143-44, 158, 169, 181, 185, 189-91, 210, 214, 243-47, 257-65
Village 37

Wire-frame models 16, 250

FREE SAMPLE REQUEST JOURNAL

Pattern Analysis and Applications

Editor-in-Chief: Sameer Singh, Department of Computer Science, University of Exeter, UK
Email: espaa@exeter.ac.uk or telephone
+44 (0) 1392 264061 or visit the
journal web site at: www.dcs.exeter.ac.uk/paa

ISSN: 1433-7541 (print)
1433-755X (electronic)
4 issues per year

Volume 3, 2000, 4 Issues
2000 Institutional Rate: £204.00
Personal Rate: £60.00

Plus carriage charges. Prices are subject to change without notice. In EU countries the local VAT is effective.

JOINT RATE FOR PAA & NCA
(Neural Computing and Applications)
AVAILABLE TO BCS SGES MEMBERS

> Free sample copy available from
> Journals Department
> Springer-Verlag London Ltd
> Sweetapple House
> Catteshall Road
> Godalming
> Surrey GU7 3DJ, UK.
> Tel: +44 (0)1483 414142
> Fax: +44 (0)1483 421270
> Email: jessica@svl.co.uk

The philosophy behind **Pattern Analysis and Applications** is to publish papers on pattern recognition and their industrial applications in a number of areas.

This research deals with the identification and development of novel pattern analysis methods which can be applied in vision, speech, robotics, multimedia, intelligent control and character recognition based on AI.

This contemporary, state of the art journal provides a much needed forum for original research in intelligent pattern analysis and applications in computer science and engineering.

Under the guidance of the editor and an international advisory board, **Pattern Analysis and Applications** offers a comprehensive study of the most current issues in the field and bridges the gap between academic research and the needs and research focuses of industry.

Now covered in Science Citation Index Expanded (also known as SciSearch®), ISI Alerting Services, the CompMath Citation Index® and Current /Contents /Engineering, Computing and Technology

Society rates available on request   Springer

To request a free copy, or to subscribe to our journals, please contact Springer-Verlag London Limited
Tel: +44 (0)1483 414142 Fax: +44 (0)1483 421270 Email: jessica@svl.co.uk